T0141712

Structural and Mediator Lipidomics

Structural and Mediator Lipidomics

A Functional View

Michel Lagarde

With contributions by

Michel Guichardant, Céline Luquain-Costaz, Nathalie Bernoud-Hubac, Catherine Calzada & Evelyne Véricel

(Lyon University, UMR Inserm 1060, UMR Inra 1397 CarMeN Laboratory, INSA-Lyon, Villeurbanne, France)

EPFL Press

A swiss academic publisher distributed by CRC Press

Taylor and Francis Group, LLC
6000 Broken Sound Parkway,
NW, Suite 300, Boca Raton,
FL 33487, USA

Distribution and Customer Service
orders@crcpress.com
www.crcpress.com

Library of Congress
Cataloging-in-Publication Data
A catalog record for this book
is available from the Library
of Congress.

The publisher and author express their thanks to the Institut National des Sciences Appliquées de Lyon (INSA-Lyon) for its generous support towards the publication of this book.

This book is published under the editorial direction of Professors Jean-Yves Cavaillé et Nicolas Freud (METIS LyonTec).

is the English-language imprint owned of the Foundation of the Presses polytechniques et universitaires romandes (PPUR). The PPUR publishes mainly works of teaching and research of the Ecole polytechnique fédérale de Lausanne (EPFL), of universities and other institutions of higher education.

Presses polytechniques
et universitaires romandes
EPFL – Rolex Learning Center
Post office box 119
CH-1015 Lausanne, Switzerland
E-mail ppur@epfl.ch
Phone 021/693 21 30
Fax 021/693 40 27
www.epflpress.org

ISBN 978-2-940222-92-6

Printed in Italy

Content

1

Introduction

Lipidomics is the part of metabolomics dedicated to lipids. It can be defined as the full molecular characterization of the lipid components in a defined system, as well as consideration of their function when available [Lagarde et al. *Biochim. Biophys. Acta*, 2003 & Spener et al. *Eur. J. Lipid Sci. Technol.*, 2003]. The name *lipidomics* was used for the first time in titles of scientific articles in 2003, and found more than four hundred times as such during the next decade, with 15% in the last year. Such a development has led experts in the field to distinguish between different subdomains, such as structural or mediator lipidomics, in an attempt to simplify the data and decrease the complexity of their interpretation. A way to keep an integrated lipidomics approach that includes the structures and functions of lipids in a particular field can be called targeted lipidomics.

In the present essay, the authors present lipid molecular species which have been well defined for their biological activities, whatever the biological system concerned, with reference to the closely related structural lipids. A special emphasis will be given to bioactive lipids issued from membrane lipids, especially polyunsaturated fatty acid derivatives produced through various oxygenation processes. The most numerous bioactive lipids have been described in the frame of cell lipid signaling, so the enzymes associated with the production of these bioactive lipids and receptor proteins relating to their action are also reported.

As methodological aspects are crucial considerations in characterizing the various lipids, the most common methods in use will be presented, together with a dynamic approach of the fluxes relating to the bioactive molecules, trying to anticipate a true fluxolipidomics. An attempt to visualize the lipids in their membrane superstructures will also be reported as imaging lipids.

The readers addressed with this essay are Master and PhD students as well as scientists starting to be interested in lipids involved in cell signaling. Also, this essay may help medical doctors interested in pathophysiological molecular bases of diseases involving lipids.

2

Bioactive lipids and their precursors

The numerous bioactive lipids require a comprehensive view of their immediate precursors and the more stable components that are usually structural components of biological membranes.

2.1 Glycerolipids

Glycerolipids result from the esterification of the three alcohol functions of glycerol by fatty acids and/or phosphate. The common glycerol backbone residue is either linked to three fatty acyls, leading to the storage lipid class called triacylglycerols (trivial name: triglycerides) or linked to two fatty acyls and one phosphate that esterifies the third alcohol, one of the two primary alcohols present in glycerol (glycero-phospholipids). The third position/alcohol of glycerol may also be derived by carbohydrates with osidic or ether linkage. There is no asymmetric carbon in glycerol, but carbon number two becomes asymmetric in most glycerolipids as the chemical groups at position one and three (acyls or polar groups) are different, except for triacylglycerols that have the same acyl group at the external positions. Natural glycerolipids are usually defined as *L* with the *S* configuration for carbon number two (Fig. 1). This stereochemistry is taken into consideration with "stereo-chemically numbered" (*sn*) preceding the number, e.g. *sn*-2 for the asymmetric carbon and *sn*-1/3 for external positions.

2.1.1 Triacylglycerols

As storage lipids, triacylglycerols (TAG) are both qualitatively and quantitatively important in plant seeds and adipose tissues of animals in the frame of further mobilization to provide energy-rich molecules. They also are substantial components of blood plasma lipoproteins, especially those called TAG-rich lipoproteins, namely chylomicrons and very low-density lipoproteins (VLDL) in which various types of lipids are non-covalently associated with specific proteins. Chylomicrons and VLDL are the privileged vehicles of TAG from the intestine and liver, respectively (see Plasma lipoproteins, pp. 87-95). When TAG are mobilized from storage tissues or blood plasma lipoproteins

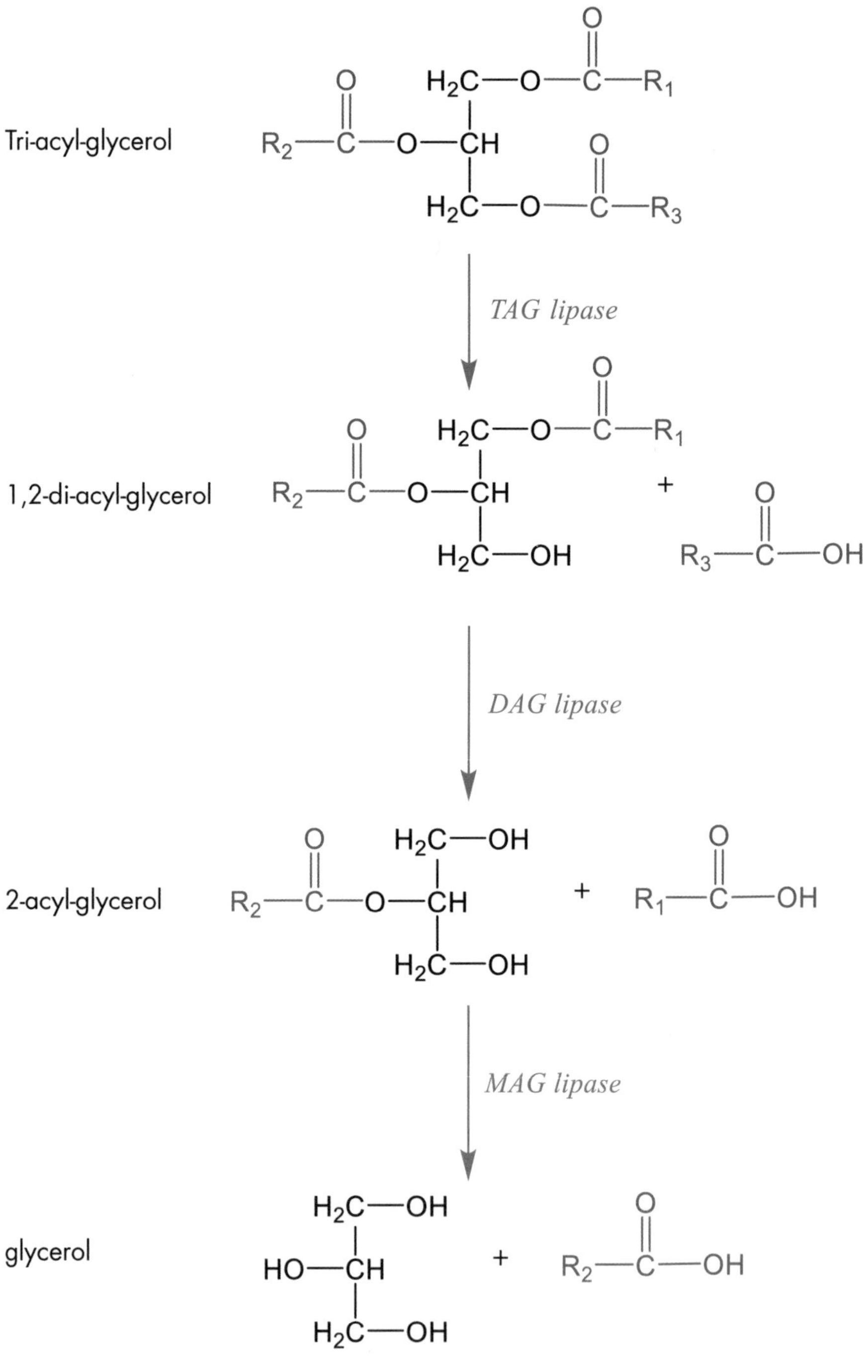

Tri-acyl-glycerol $R_2-C(=O)-O-CH(CH_2-O-C(=O)-R_1)(CH_2-O-C(=O)-R_3)$

TAG lipase

1,2-di-acyl-glycerol $R_2-C(=O)-O-CH(CH_2-O-C(=O)-R_1)(CH_2-OH)$ + $R_3-C(=O)-OH$

DAG lipase

2-acyl-glycerol $R_2-C(=O)-O-CH(CH_2-OH)(CH_2-OH)$ + $R_1-C(=O)-OH$

MAG lipase

glycerol $HO-CH(CH_2-OH)(CH_2-OH)$ + $R_2-C(=O)-OH$

Figure 1 Triacylglycerols and their cleavage by lipases.

by TAG lipases, acyl residues at the *sn-1* or *sn-3* positions are hydrolyzed to release fatty acids (Fig. 1), that become available to other tissue via albumin transport within the blood stream. Saturated and mono-unsaturated fatty acids are mainly used for energy supply through beta-oxidation, but polyunsaturated fatty acids may contribute to cell signaling as described below in Fatty acids and derivatives, p. 20-51.

Tri-acyl-glycerols (TAG) result from the esterification of the three alcohol groups of glycerol by different fatty acids. TAG are then very hydrophobic compared to their components, especially glycerol that is water soluble. The molecular species of TAG are numerous due to the various possible acyl residues R_1, R_2, and R_3. In some rare cases the three fatty acyls are the same, providing a symmetrical TAG, e.g. tripalmitin if each fatty acid involved in the esterification was palmitic acid (16:0). When R_1 differs from R_3, carbon 2 of the glycerol moiety becomes asymmetric. The natural TAG are of the L or S configuration, and the corresponding carbon is called *sn*-2 for "stereo-chemically numbered 2". By extension, the external carbons of the glycerol moiety are called *sn*-1 and *sn*-3 from top to down in the figure.

TAG are hydrolyzed by lipases, the first being called TAG lipase, cleaves the *sn*-3 ester. The resulting di-acyl-glycerol (DAG) is cleaved by DAG lipase that hydrolyses the *sn*-1 position. Finally, mono-acyl-glycerol (MAG) is further hydrolyzed by MAG lipase into glycerol plus R_2-COOH.

2.1.2 Glycerophospholipids

They are the most abundant lipids in biological membranes in general, except in thylakoids of plant cells where galacto-lipids are major lipid components (see Galacto-lipids, pp. 52-53). Glycerophospholipids differ both in terms of polar heads at the *sn*-3 position, and fatty acyl moieties at the *sn*-1 and *sn*-2 positions. Having two fatty acyl residues per molecule, the number of phospholipid molecular species is quite high. In addition, ether phospholipids are glycerol-phospholipids that have an alkyl (saturated ether) or alkenyl (alpha unsaturated ether) chain at the *sn*-1 position. The general structure is described in Figure 2.

The group of diacylglycerophospho-X, with X corresponding to the residue of X-OH (e.g. X-OH: ethanolamine: HO-CH_2-CH_2-NH_2; X: -CH_2-CH_2-NH_2) after esterification by the phosphate group that also esterifies the *sn*-3 primary alcohol of glycerol, is the most abundant lipid form in biological membranes, notably in animal cells.

2.1.2.1 Classes, subclasses and molecular species of glycerophospholipids

The different classes of glycerophospholipids relate to X-OH, which can be: choline, ethanolamine, L-serine, myo-inositol (possibly further phosphorylated

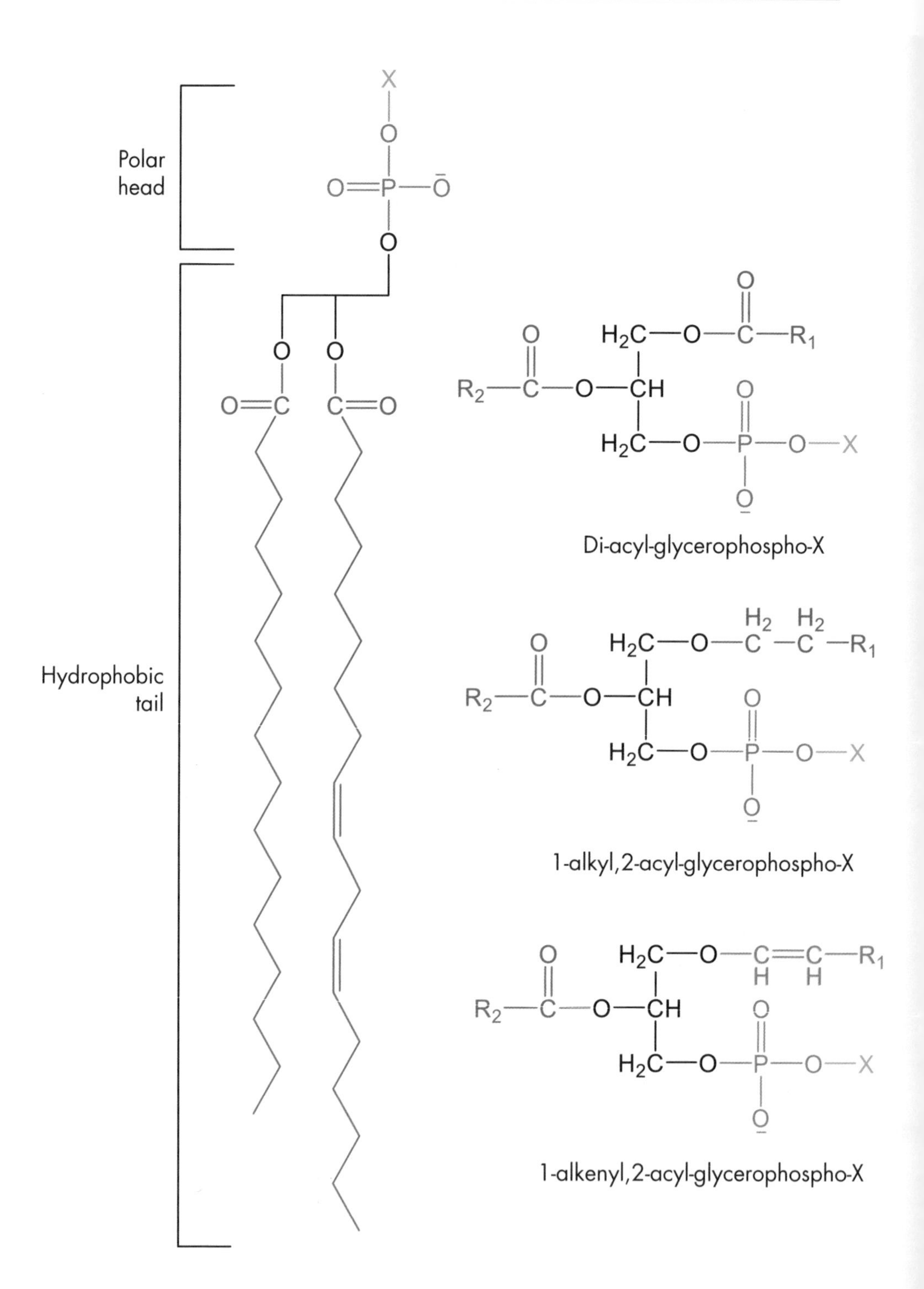

Figure 2 General structure of glycerophospholipids.

Often abbreviated GPL, glycero-phospholipids have two acyl residues at the *sn*-1 and *sn*-2 positions, with the *sn*-3 alcohol of the glycerol moiety esterified by phosphoric acid. Only one GPL (phosphatidic acid or phosphatidate) has a free phosphate group. Most GPL are phosphodiesters with another alcohol esterified by one of the remaining acidic group of the phosphoric acid moiety (see Fig. 3). In Figure 2, X represents the second ester of the phosphate group with X-OH for the initial alcohol (see Fig. 3). The phosphate-X moiety is called the polar head of phospholipids whereas the counter part with the glycerol residue esterified twice is called the hydrophobic tail. In biological membranes, the polar head interacts with the water phases (outside as well as inside the cell) and the hydrophobic tail interacts with those from other phospholipids to make the hydrophobic core of the membranes.

The di-acyl species of glycerol-phospholipids (di-acyl-GPL) are the most frequent ones, but some are ethers instead of esters at the *sn*-1 position. They are called alkyl,acyl- and alkenyl,acyl-GPL. In the latter case, there is a cis/Z unsaturation next to the ether bond.

Whereas the *sn*-1 position has a saturated acyl or alkyl residue, the *sn*-2 position is always unsaturated and very frequently polyunsaturated. There is one well-known exception that is dipalmitoyl-glycero-phosphocholine as an important component of pulmonary surfactant.

Figure 2 represents the chemical structure of 1-palmitoyl,2-linoleoyl-GPX.

once, twice, or three times), or glycerol (Fig. 3). X is then the only difference in the polar head group: phosphate-X. The corresponding glycerophospholipids are called: phosphatidylcholine (PC) or choline glycero-phospholipids because they may contain ether phospholipids, especially as an alkyl form (see below), phosphatidylethanolamine (PE) or ethanolamine glycerophospholipids as they may also contain an ether moiety, especially as an alkenyl or alpha-unsaturated alkyl form, and phosphatidylserine (PS). Glycerophospholipids also comprise phosphoinositides (PIs) that include the most common one called phosphatidylinositol (PI), and the further phosphorylated derivatives phosphatidylinositol-4-phosphate (PIP) and phosphatidylinositol-4,5-bisphosphate (PIP_2). A further phosphorylation on carbon 3 is possible in response to cell activation, leading to phosphatidylinositol-3,4,5-trisphosphate (PIP_3). In contrast to the other phosphoinositides, PIP_3 is a quite transient component of the membranes. Phosphatidylglycerol (PG) is also present, especially in prokaryotic membranes.

The relative abundance of these glycerophospholipids in animal cells varies according to the different membranes of the cell and even according to the different cell types. It can be stated that PC is predominant, followed by PE, then PS and finally PIs.

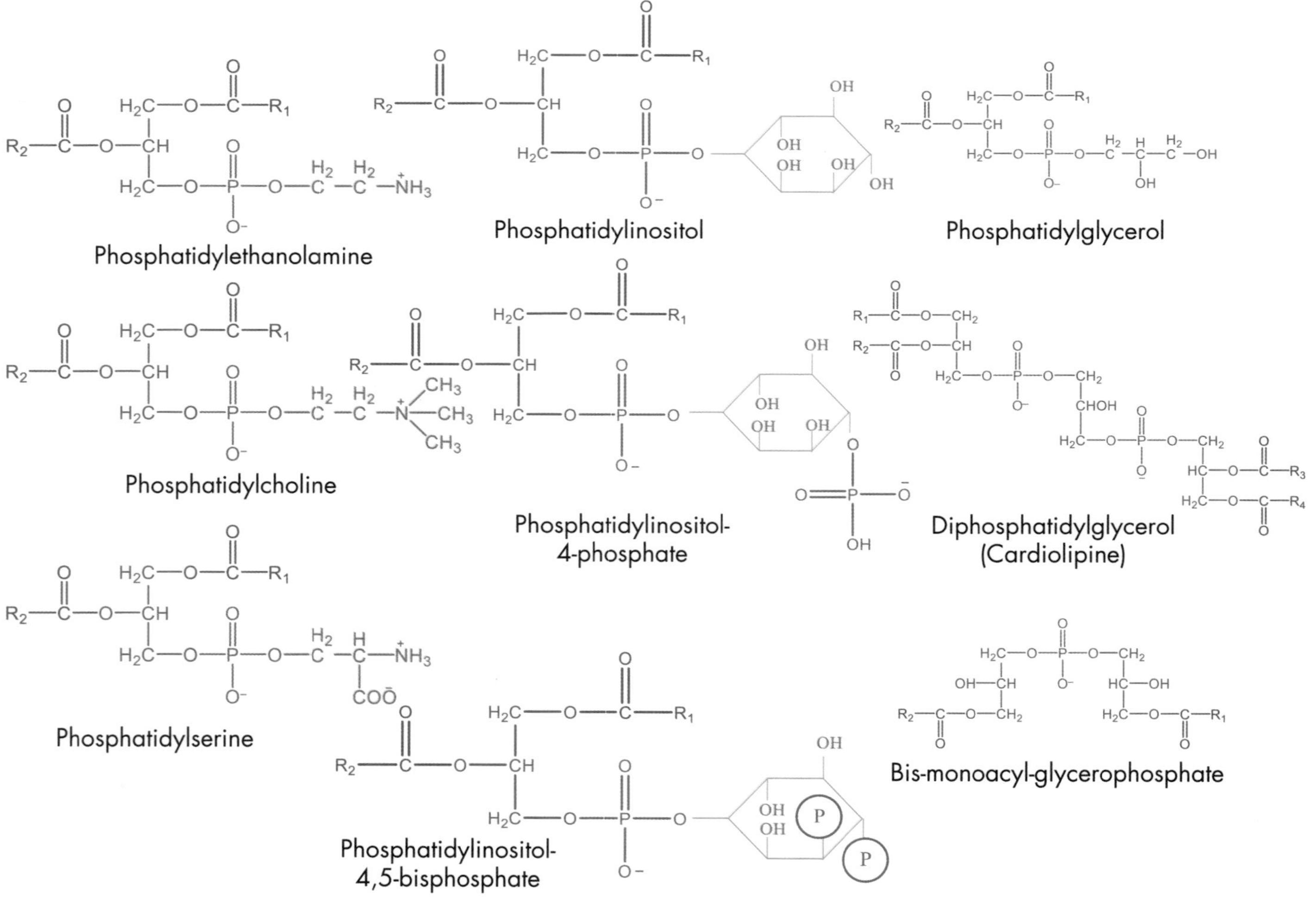

Figure 3 Different classes of glycerophospholipids according to their polar head group.

Not only GPL differ on the basis of the ester *versus* ether residue at the *sn*-1 position, and according to the nature of the different acyls at the *sn*-2 position as well as the *sn*-1 position in diacyl-GPL, but mainly vary according to the nature of X. Most frequently, X-OH are ethanolamine, or N-trimethyl-ethanolamine/choline, or carboxy-ethanolamine/serine. The corresponding GPX are phosphatidyl-ethanolamine (PE), -choline (PC) and -serine (PS), respectively. These GPL classes represent almost 90% of the cell membrane GPL in animals.

Other minor (around 10% altogether) but functionally important GPL are called phospho-inositides. Their X moiety is myo-inositol in phosphatidyl-inositol (PI), the main phosphoinositide, further phosphorylated into PI-4-phosphate (PIP) and further into PI-4,5-bisphosphate (PIP_2). For reasons of hindrance, the two phosphates groups in PIP_2 are shown as a "P" with a circle.

Finally, more complex or specific GPL are phosphatidyl-glycerol (PG), that is quantitatively important in prokaryotes, but rather an intermediate in eukaryotes, and diphosphatidyl-glycerol (DPG) also called cardiolipin (CL) because of its prominence in mitochondria. CL are almost exclusively located in the inner membranes of mitochondria which are especially numerous in cardiomyocytes. Cardiolipin are twice as big as the other GPL with two phosphate groups and four acyl residues. This important GPL class in the inner membrane of mitochondria is believed to be responsible for at least part of the well-known impermeability of this membrane. The third specific GPL is bis-monoacyl-glycerol-phosphate (BMP), a position isomer of PG, which is quantitatively minor, and located in cell late endosomes. These three GPL have in common to contain a second glycerol moiety as the X group. BMP is a position isomer of PG with each glycerol moiety being esterified once.

Among PIs, PIP and PIP_2 are less present, reaching around 10% of total PIs. This class of phospholipid is then quantitatively minor, although it is considered as a major bioactive lipid in cell signaling in several ways. Among PIs, it is worth to point out the very minor component PIP_3, which does not accumulate, as already mentioned above, but rises in response to activation of some tyrosine kinase receptors such as that of insulin, resulting in PI-3-kinase activation. PI-3-kinase is especially active upon PIP_2, giving rise to PIP_3 which then acts as a messenger.

Finally, it is also worth to mention two peculiar glycero-phospholipids that contain two acylated glycerol moieties. They are bis-phosphatidyl-glycerol or cardiolipin (CL) and bis-monoacyl-glycerophosphate (BMP), a position isomer of PG.

- CL is especially rich in unsaturated fatty acyls (often at each *sn* position), mostly linoleoyls. CL is almost exclusively located within the inner membrane of mitochondria, where it plays an important structural role for respiratory chain cytochrome insertion. The "double" size of CL, compared to classic glycerophospholipids, is also believed to be responsible for the relative impermeability of the mitochondrial inner membrane.
- BMP is a quantitatively minor phospholipid (the most minor one) produced from PG by a complex transacylation process. BMP is quite specifically located in the luminal membranes of late endosomes where it seems to regulate the cholesterol homeostasis and glycosphingolipid degradation.

It is usually stated that the *sn*-1 position of glycerophospholipids is occupied by a saturated acyl whereas there is a cis/Z unsaturated (often polyunsaturated, meaning with at least two double bonds) acyl at the *sn*-2 position. This is crucial for the fluidity of cell membranes, because of the folding in the sn-2 chain due to its cis/Z geometry. Then the hydrophobic part of those glycerophospholipids takes more space than the polar head, allowing glycerophospholipids to move relative to each other within one leaflet of the membrane. There is an important exception in lung surfactant where dipalmitoyl-glycero-phosphocholine (palmitoyl is devoid of double bond) is the major lipid.

As already mentioned above, two classes of glycerophospholipids, PC and PE, can have an alkyl or alkenyl chain at the *sn*-1 position instead of an acyl chain. In case of the alkenyl chain, the phospholipid subclass is called plasmalogens. These phospholipids correspond to two subclasses belonging to the so-called ether phospholipids. The alkyl chain is preferentially present in choline phospholipids, whereas the alkenyl chain is predominant in ethanolamine phospholipids. In some cells, the later subclass may reach 40% of the total ethanolamine phospholipid class. Of note, choline plasmalogens are found in substantial amounts in cardiac cells.

The biological role of ether phospholipids compared to the abundant diacyl-glycero-phospholipids is not completely understood. Whereas the latters are recognized for both their structural and precursor roles, that of the formers seems to be more specific. As a matter of fact, 1-alkyl,2-arachidonoyl-glycerophosphocholine (-GPC) is the precursor of Platelet-Activating Factor (PAF) that is generated in some cells in response to specific stimuli. This precursor is cleaved by $cPLA_2$ (see below in Lipolytic enzymes involved in the release of bioactive lipids, pp. 11-18) to release 1-alkyl-GPC that is re-acylated into 1-alkyl,2-acetyl-GPC or PAF by an acetyltransferase using acetyl-CoA as donor. PAF is a highly bioactive phospholipid, relatively water soluble, due to its short chain at the *sn*-2 position, it is able to aggregate platelets, its primary function, induce inflammation by promoting blood plasma exudation, and bronchoconstriction.

On the other hand, 1-alkenyl,2-acyl-glycerophosphoethanolamine as well as 1-alkenyl,2-acyl-glycerophosphocholine have been described as endogenous antioxidant, because of the ability of vinyl ether in the alkenyl moiety to trap oxygen radicals (Fig. 4). As a result, reactive aldehydes are produced that may be deleterious molecules, as described in Other peroxidation products, pp. 44-51.

Since glycerophospholipids have a dual and variable hydrophobic chain at positions *sn*-1 and *sn*-2, the molecular species within one class and even one subclass of glycerophospholipid are quite numerous, and this is relevant to different biological functions. This biological diversity requires the analytical high-performance liquid chromatography coupled with mass spectrometry (LC-MS) that will be detailed in Lipidomics approaches, pp. 80-81.

2.1.2.2 Lipolytic enzymes involved in the release of bioactive lipids

Glycerophospholipids that self-assemble to make the membrane bilayer are specifically cleaved to generate bioactive lipids or substrates for further metabolism, including pathways to produce bioactive lipids.

The cleaving enzymes are called phospholipases and belong to four different types, according to the cleavage site on glycerophospholipids (Fig. 5 & 6).

Phospholipase A_1 (PLA_1) hydrolyses the ester bond at position *sn*-1, releasing R_1-COOH and the counter molecule 2-acyl-lysophospholipid (Fig. 5). Ether phospholipids are not cleaved by PLA_1. PLA_1 are believed to play a main role in digestion to degrade dietary phospholipids, with no obvious relationship with lipid signaling. As a matter of fact, it is worth to state that degrading glycerophospholipids by removing the fatty acyl residue at the *sn*-1 position can easily be made by most of triacylglycerol lipases, making glycerophospholipids possible substrates of those lipases. The relative instability of 1-lyso,2-acyl-glycerophospholipids, with spontaneous migration of the acyl group to position *sn*-1, leads to the stable 1-acyl,2-lyso-glycerophospholipids. This is due to the higher reactivity of the primary alcohol at position *sn*-1 compared to the secondary alcohol at position *sn*-2 in the equilibrium between the two position isomers.

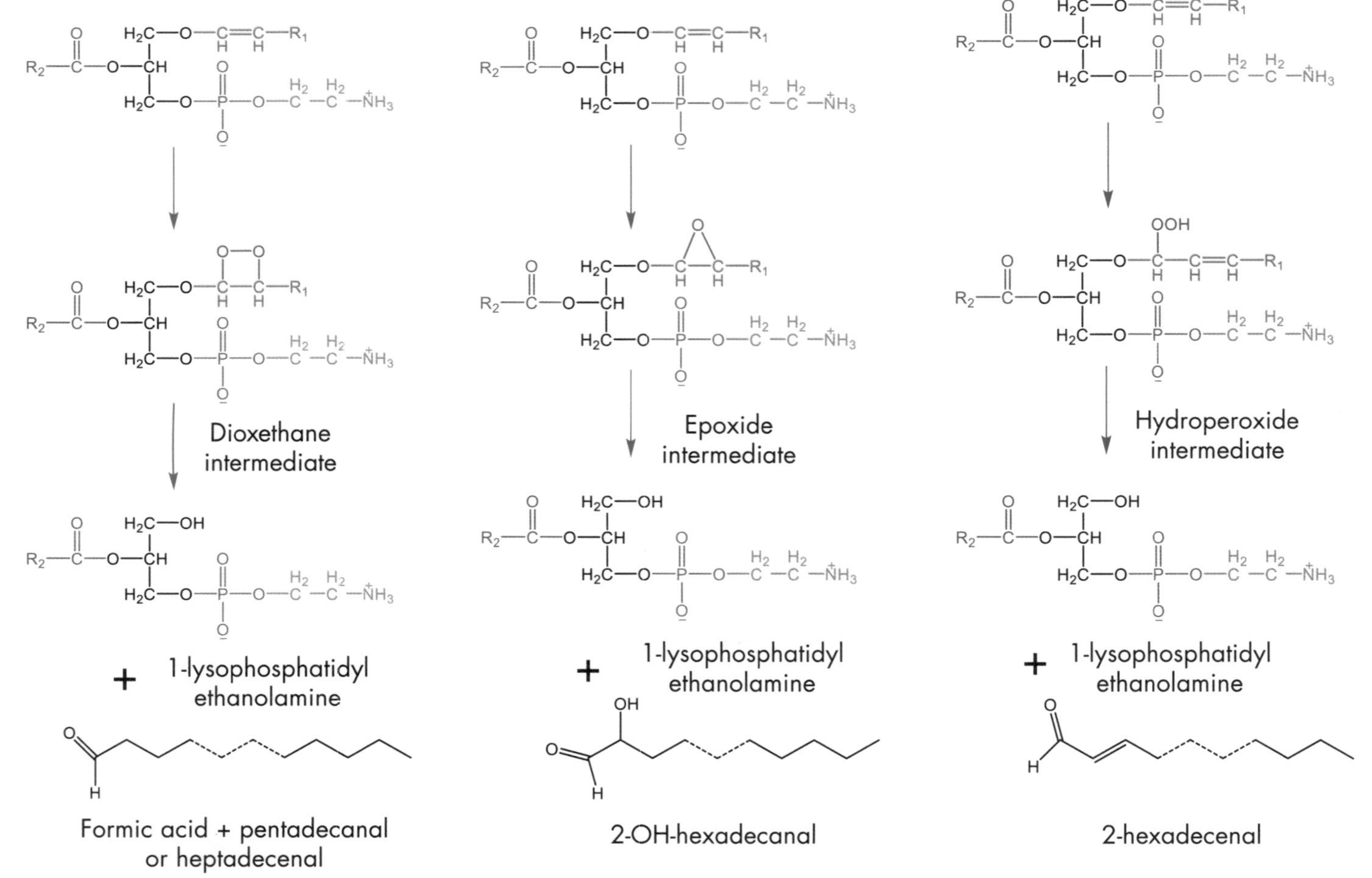

Figure 4 Scheme of possible cleavage of the alkenyl,acyl residue by oxygen radicals.

Alkenyl,acyl-GPX, mainly alkenyl,acyl-GPE called plasmalogens, are known as antioxidants by trapping reactive oxygen species (ROS) due to vinyl-ether reactivity (because of the double bond next to the oxygen atom at the *sn*-1 position).

At least three different mechanisms are known for the ROS-induced cleavage of the alkenyl moiety. They all lead to formation of aldehydes, the alpha-unsaturated aldehyde (on the right of the figure) being the most reactive and deleterious. It can be noticed that in all cases the primary alcohol at the *sn*-1 position of the initial GPL is re-generated as a lyso-GPL, allowing the *sn*-2 acyl (highly unsaturated in plasmalogens) to migrate to the *sn*-1 position.

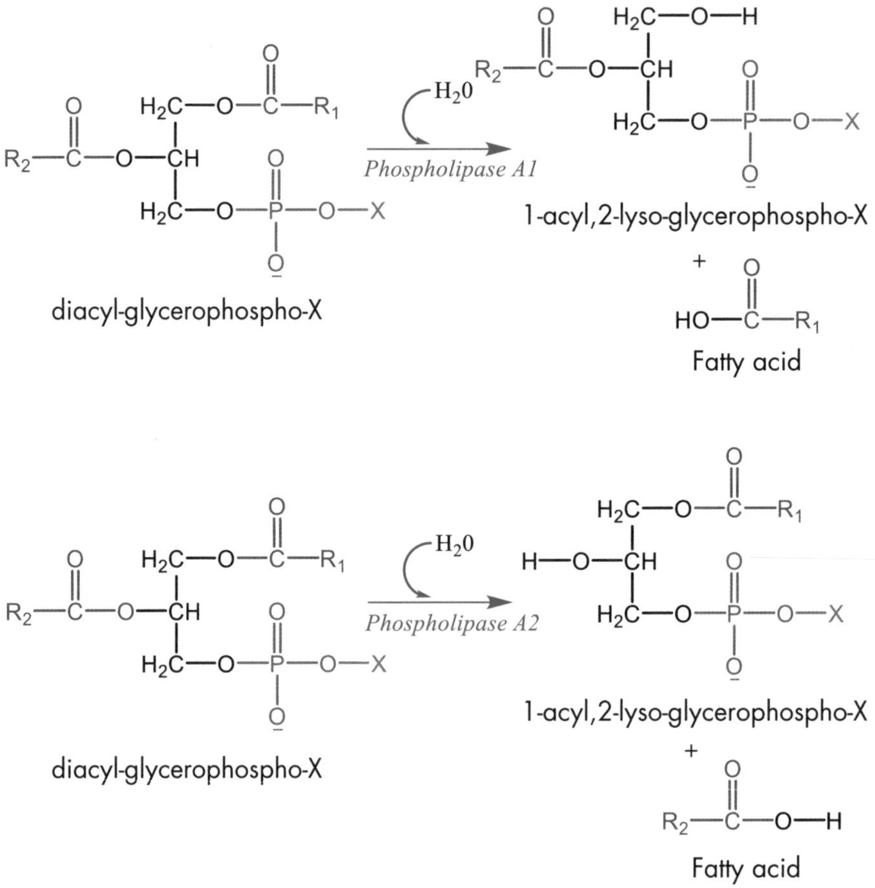

diacyl-glycerophospho-X → (H_20, *Phospholipase A1*) → 1-acyl,2-lyso-glycerophospho-X + Fatty acid

diacyl-glycerophospho-X → (H_20, *Phospholipase A2*) → 1-acyl,2-lyso-glycerophospho-X + Fatty acid

Figure 5 Cleavage of glycero-phospholipids by phospholipases A_1/A_2 and their products.

This figure shows the removal of the acyl residues by phospholipases A (PLA). PLA_1 hydrolyzes the *sn*-1 and PLA_2 hydrolyzes the *sn*-2 positions. The products are the corresponding fatty acids and resulting lyso-GPL. As mentioned in Figure 4 legend, 1-lyso,2-acyl-GPL is rather unstable due to the reactivity of primary alcohol at the *sn*-1 position, allowing the acyl at the *sn*-2 position to migrate to the *sn*-1. So, the lyso derivatives produced from both PLA_1 and PLA_2 hydrolyses are likely to be finally 1-acyl,2-lyso-GPL, with a saturated lyso compound from PLA_2 and unsaturated lyso compound from PLA_1 actions.

A specific lysophospholipase D acting upon lysoPC produced by cleavage of PC by PLA_2 is called autotaxin. It cleaves lysoPC to release lysoPA, a potent mitogenic agent that is relatively water soluble compared to the initial PC. LysoPA may then migrate from the site of its production to specific membrane receptor on target cells.

Phospholipases A_2 (PLA_2), which cleave the ester bond at position *sn*-2 of glycerophospholipids (Fig. 5), are clearly more relevant from the signaling viewpoint as they release unsaturated fatty acids (essentially polyunsaturated fatty acids) that can be further oxygenated into highly bioactive products. Regarding their molecular biology, they can be subdivided into three categories: secreted ($sPLA_2$), cellular ($cPLA_2$), and calcium-insensitive ($iPLA_2$).

- $sPLA_2$ have been first described. They include pancreatic PLA_2 acting in digestion of dietary phospholipids, venom PLA_2 from snakes and bees, and inflammatory cells PLA_2 to participate in the inflammation process. Their common feature is their low molecular mass (14 kDa) and the requirement for high concentrations of calcium ions (mM), which is relevant to the extracellular concentration of this positive ion. These PLA_2 do not exhibit fatty acyl specificity at the *sn-2* position, and cleave all glycerophospholipid classes.

- $cPLA_2$ have been the object of much attention because they play a crucial role in releasing polyunsaturated fatty acids from the *sn-2* position, especially arachidonic acid for which they exhibit a rather high specificity. In contrast to $sPLA_2$, $cPLA_2$ have a high molecular mass (85 kDa) with a big change in electrophoretic mobility in SDS-PAGE (to evaluate the molecular mass of proteins) due to the phosphorylation of up to two serine residues, in response to cell activation. This phosphorylation has a drastic effect on the 3D conformation of the enzyme, which explains the big change in electrophoretic mobility. This is required for phospholipid hydrolytic activity. Such a phosphorylation results from a cascade of kinase activation, very often of various MAPK (Mitogen-Activated Protein Kinases), some are activated by oxidative stress, and others by calcium ions. Of note, $cPLA_2$ require micromolar concentrations of calcium ions, which is the situation found in the cytosol of activated cells. This means that delocalization of calcium ions from the endoplasmic reticulum usually occurs upstream of the $cPLA_2$ activation.

- $iPLA_2$ are not so well characterized. They require only nanomolar concentrations of calcium ions, which is the situation found in the cytosol of non-activated cells. Indeed $iPLA_2$ are believed to play a role in fatty acyl remodeling, allowing cell membrane phospholipids to change their *sn-2* fatty acyl composition according to the environment. In activated cells, some studies have indicated that $iPLA_2$ could be involved in the release of arachidonic acid from ethanolamine plasmalogens.

Phospholipases C (PLC) cleave the phosphodiester bond between the glyceride moiety and the polar head (Fig. 6), producing a water soluble entity, mainly phospho-choline and phospho-inositol (phosphorylated or not), and a more hydrophobic product compared to the substrate, di-acyl-glycerol (DAG), which remains within the membrane and even tends to sink into it.

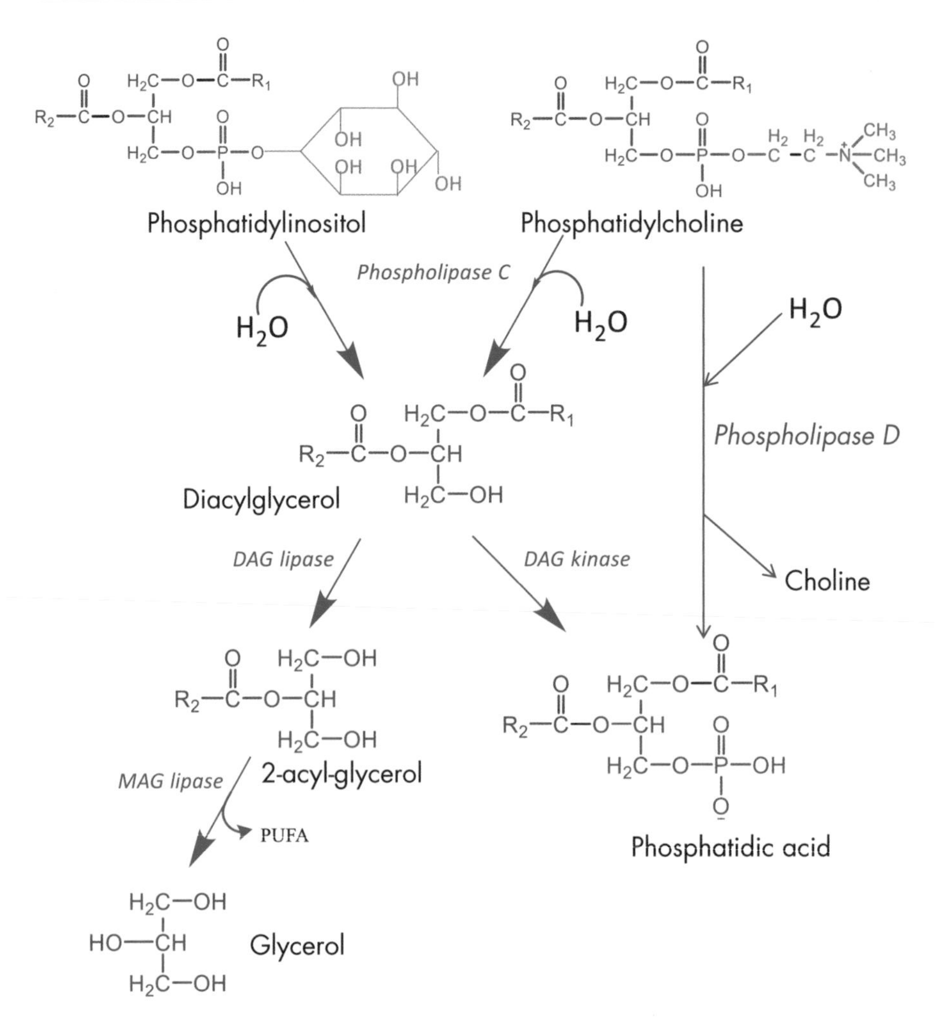

Figure 6 Phospholipid cleavage by phosholipases C & D, and metabolic fate of DAGs through DAG/MAG lipases and DAG kinase.

Two categories of PLC are commonly distinguished according to their activation process.

- PLC_β is activated by cell agonists acting through hetero-trimeric G-proteins coupled with their specific receptors. The G-protein involved here is called Gq.
- PLC_γ is activated by self-phosphorylated tyrosine-kinase receptors (such as the insulin receptor) which bind PLC when those receptors are highly phosphorylated, and consequently make it active.

Both enzymes are calcium-ion sensitive and act upon PC and PIs (Fig. 6), with a preferential hydrolysis of PIs by PLC_γ. Out of the common product DAG that has a characteristic bioactivity (see below), the biological activities of

This figure shows the cleavage of GPL by two phosphodiesterases, PLC and PLD. The lipid products are 1,2-DAG and phosphatidic acid (PA), respectively. Stability of 1,2-DAG is questionable as it might isomerize into 1,3-DAG because of the high reactivity of the primary alcohol at the *sn*-3 position, although such an isomerization is not facilitated by a phosphate group as in 1-lyso-2-acyl-GPL. PA is the simplest GPL with X equals H. PA has been described as a substrate of PLA, especially PLA_2.

The hydrolysis of phosphoinositides (PIs), represented by PI in the figure, as well as from PC, and the metabolic fate of DAG are represented. As shown in Figure 1, DAG may be further hydrolyzed by the DAG/MAG lipases cascade or phosphorylated into PA by DAG kinase. Therefore, this second route may be connected with the PLD pathway that more specifically hydrolyzes PC in animals, and PE in plants to produce PA. PA may also be hydrolyzed into DAG by PA-phospho-hydrolase, making PLC and PLD highly connected.

PIP_2 is a well-known substrate of PLC to produce DAG and IP_3 that is a potent ionophoric agent to release calcium ions stored within the endoplasmic reticulum (ER). PIP_2 being almost exclusively located on the cytoplasmic leaflets of animal membranes, the high water solubility of IP_3 allows it to easily reach the ER. On the other hand, the molecular species of PIs being principally the palmitoyl,arachidonoyl species, PA produced through the PLC/DAG kinase pathway has substantially different acyl moieties from that issued the PC/ PLD pathway, as the molecular species of PC are much more diverse than that of PIs.

polar heads have not been described as such, except for the hydrolysis product of PIP_2, inositol 1,4,5-trisphosphate (IP_3). PIs are mainly located in the inner leaflet of plasma membranes and the cytosolic leaflet of nuclear membranes, so that the polar heads are released into the cytosol. In case of PIP_2 hydrolysis, the highly water-soluble product IP_3 acts to release calcium ions from the endoplasmic reticulum through specific receptors on this organelle. It is often proposed that the activation of PLCs, especially PLC_γ, precedes that of $cPLA_2$ to yield calcium ions for the activation of the latter.

Phospholipases D (PLD) are also glycero-phospholipid phosphodiesterases that cleave the phosphor-ester bond between the phosphate group and the alcohol residue of the polar head (Fig. 6). They are mainly active upon PC, producing the most simple glycerophospholipid, phosphatidate (PA), and choline which can be recycled for phospholipid synthesis and/or production of the neurotransmitter acetylcholine, depending on the environment. The biological relevance of PA will be detailed below.

- Oleic acid sensitive PLD has been described as activated by fatty acids, especially oleic acid. However, this type of PLD is not much mentioned nowadays, so will not be taken into consideration in this presentation.

- PIP_2-sensitive PLD comprises two subtypes according to the requirement of small G-proteins for their activation, and called PLD_1 and PLD_2. Both require PIP_2 as a cofactor (it is worth stating that PIP_2 is not a substrate of PLD) with binding of PLD to the negatively charged PIP_2 which is essentially located at the inner leaflet of plasma membranes and cytosolic leaflet of the nuclear membrane, as mentioned above. PLD_2 is less sensitive to regulation than PLD_1, and is highly sensitive to the small G-proteins GTPases Rho and ARF (ADP-ribosylation factor). PLD_2 is also regulated by protein kinase C and the H-Ras cascade.

2.1.2.3 Released bioactive lipids

Released bioactive lipids include fatty acids, mainly PUFA, and more or less hydrophobic entities that remain either embedded within the membrane or leave it to act within the producer cell or in the vicinity, if they are water soluble enough. Although those entities belong to lipids, they may be partly water soluble, especially at low concentrations which are often fully active. Their local focusing onto specific receptors could also make them active, even at low concentrations.

• **1** Di-acyl-glycerol, phosphatidate and lyso-phospholipids

Di-acyl-glycerols (DAGs) can be produced by cleavage of PC and/or PIs by PLC, or by dephosphorylation of PA issued from the PC cleavage by PLD (Fig. 6). The latter pathway requires the PA phospho-hydrolase that seems to be quite ubiquitous. The resulting DAGs have diverse fatty acyl composition (diverse molecular species), depending on the primary source: PC or PIs. From PIs, meaning exclusively through PLC activation, the major species is 1-stearoyl,2-arachidonoyl-glycerol, but from PC (through both PLC and PLD activation), the latter species is one among others such as 1-palmitoyl,2-linoleoyl-glycerol which is the most prominent in number of cell PC.

The most relevant biological activity of DAGs is to activate protein kinases C (PKCs). However, the numerous PKC isoforms may not respond the same way to the different DAGs, the *sn-2* PUFA-containing DAGs being more active, whereas the alkyl,acyl-glycerol (issued from the cleavage of ether PC by PLC) may oppose the activation by DAGs.

DAGs may be further cleaved by DAG lipase giving rise to *sn-2* monoacyl-glycerol (MAGs), ultimately hydrolyzed into glycerol plus PUFA (as this is the major fatty acyl moiety at the *sn-2* position of DAGs) by MAG lipase (Fig. 6). Among *sn-2* monoacyl-glycerols produced, 2-arachidonoyl-glycerol (2-AG) is of special interest as a source of arachidonic acid for further metabolism (see Fatty acids and derivatives, p. 20), or as a bioactive lipid itself with cannabinoid activity (see Endocannabinoids, pp. 57-59).

A second possibility for DAG metabolism is the phosphorylation into phosphatidate (PA) by DAG kinase, an ubiquitous enzyme. The PA obtained may be different from that produced by PLD cleavage of PC, depending on the origin of DAG (Fig. 6), as stated above for molecular species of DAGs whether they directly derive from PIs cleavage, or from PC with further dephosphorylation of PA.

As already mentioned, PA may be produced by two pathways: through the direct cleavage of PC by PLD, and less directly through the hydrolysis of PC and/or PIs followed by the phosphorylation of the products (DAGs) by DAG kinase. The normal metabolic fate of PA is its hydrolysis back to DAGs by PA phospho-hydrolase. Thus, the metabolisms of DAGs and PAs are clearly entangled.

In terms of biological activity, PA has multiple targets, going from the activation of NADPH oxidase, the main producer of superoxide anions, to the activation of some isoforms of nucleotide phosphodiesterases. The activation of protein kinase C by PA has even been reported, but it is hard to tell whether this is an effect of PA or of the DAG produced by cleavage of PA.

Lyso-phospholipids (LPLs), more precisely lyso-glycero-phospholipids, are the primary products from glycero-phospholipids cleaved by PLA_2 (even PLA_1) (see Fig. 5). LPLs are well-known detergents, but cells are usually very active in re-acylating them, by using diverse acyl-CoAs as substrates, to make back glycero-phospholipids that are re-inserted into membranes. Their life span is then limited. However, two major LPLs can be pointed out for their biological relevance, both quantitatively and qualitatively. They are LPA and LPC, with the possibility of producing the former from the latter by the action of a lyso-phospholipase D releasing choline. This lyso-phospholipase D is also called autotaxin, and is believed to play a relevant role in the generation of LPA promoting cancer cell proliferation and metastasis through specific receptors (see below).

As a main product from PLA_2 cleavage, LPA is likely to contain saturated fatty acyl at the *sn-1* position. Only LPA issued from PLA_1 cleavage of PA, if it occurs, should have PUFA at the *sn-1* position after spontaneous isomerization/migration from the *sn-2* position.

The same considerations apply to LPC in terms of fatty acyl composition for the same metabolic reasons. In addition, it is worth stating that a substantial amount of saturated LPC is generated in blood plasma within high-density lipoproteins (HDL) by the so-called lecithin cholesterol acyl transferase (LCAT) enzyme attached to HDL which transfers the *sn-2* fatty acyl from PC to cholesterol, then producing the important lipid class called cholesterol-esters in plasma lipoproteins (see Functional dynamics, pp. 90-95). This LPC can then act in the circulation, even if it is believed to bind plasma albumin.

Biological activity of LPLs:

- LPA has a mitogenic activity, initially characterized as associated with endothelial differentiation gene (EDG) specific receptors. The EDG receptors are members of a family of eight G-protein coupled with trans-membrane receptors, three for LPA and five for another lipid mitogen: sphingosine-1-phosphate (see Sphingolipids, pp. 59-64).

- LPC has been well studied in relation to atherosclerosis and inflammatory status. LPC is recognized as responsible for alterations of endothelial-dependent vasorelaxation, endothelium permeability, as well as smooth muscle cells, monocytes/macrophages, and T-cells. These actions occur via second messengers through specific G-proteins.

• **2** Fatty acids and derivatives

Fatty acids that are released in response to phospholipid cleavage by PLA_1 are mostly saturated fatty acids, which are mainly of energetic interest due to degradation by beta-oxidation, but are not bioactive by themselves. In contrast, unsaturated fatty acids, especially polyunsaturated ones (i.e. having at least two cis/Z double bonds), are released by PLA_2 or a cascade of lipases (see above). These polyunsaturated fatty acids, having usually non-conjugated double bonds, are very often oxygenated by specific oxygenases into a great variety of bioactive products, or autoxidized into products that may be degraded into deleterious by-products.

○ Polyunsaturated fatty acids (PUFA)

The usual methylene-interrupted double bonds PUFA have at least two cis/Z double bonds, with the highest unsaturated PUFA having six double bonds. Although the chemical official nomenclature numbers the first double bond from the carboxylic carbon, they are generally separated into at least two families according to the position of the double bond closest to the omega (ω) carbon or carbon number n (Cn:x ω/n-y with n: number of carbons; x: number of double bonds; y: position of the double bond closest to the methyl end). This separation is in accordance with the two essential/indispensable fatty acids in mammals: linoleic (18:2n-6) and alpha-linolenic (18:3n-3) acids, with eighteen carbons, and two and three double bonds, the distal one at omega 6 and 3, respectively (Fig. 7). From the two essential PUFA (18:2n-6 and 18:3n-3) made in plants, animal organisms elongate and desaturate them (mostly in the liver) into twenty and twenty-two carbon products with up to six (all cis/Z) double bonds (Fig. 7).

In addition to this simple scheme of alternate desaturations and elongations from the indispensable PUFA, direct elongations may occur to produce

20:2n-6 and 20:3n-3, respectively. These two elongated products are present in mammals in measurable amounts. Even, successive elongations occur from 22:6n-3 in the retina and from 22:5n-6 in testes until the very long-chain PUFA 36:6n-3 and 32:5n-6, respectively.

Other non-essential PUFA series belong to the n-9 family. They derive from oleic acid (18:1n-9), with the rise of 20:3n-9 (Fig. 7) as a marker of linoleic acid deficiency, to mimic the conformation, but not the biological activity, of arachidonic acid (20:4n-6). Indeed, both 20:3n-9 and 20:4n-6 have the same first three double bonds in common at positions 5,8,11, then exhibiting the same folding from the carboxylic end when esterified in glycero-lipids at the *sn-2* position, which may partly mimic the three-dimensions of the glycero-phospholipid in biological membranes.

Some exceptions can be pointed out for conjugated linoleic acids (CLA) that have a cis-*trans*/*Z,E* conjugated diene geometry. They are 9*Z*,11*E*-18:2 and 10*E*,12*Z*-18:2. The former is also called rumenic acid because it is found in ruminant milk, as produced by bacteria in the rumen. The latter is issued from 18:2n-6 by isomerization due to technological processes. Rumenic acid has been reported as a useful nutrient against adiposity in the metabolic syndrome, but its isomer (10*E*,12*Z*-18:2) seems to be rather deleterious towards these metabolic effects.

Some others have three double bonds with different geometry, e.g. punicic acid from pomegranate seed oil. This is the conjugated triene (9*Z*,11*E*,13*Z*-18:3), a geometric isomer of α-eleostearic acid (9*Z*,11*E*,13*E*-18:3) from tung oil, and they have 9-*trans* isomers: β-eleostearic acid (9E,11E,13E-18:3), and catalpic acid (9E,11E,13Z-18:3) from Catalpa bignonioides seeds.

One conjugated tetraene issued from Atuna seeds is parinaric acid (9Z,11E,13E,15Z-18:4) which is fluorescent and has been used as a membrane fluorescent probe when acylating the *sn-2* position of glycero-phospholipids. Indeed, its release by phospholipases A_2 can be directly followed by fluorescence.

Although most PUFA exert specific biological activities through their oxygenated products, some of them are bioactive on their own. As a matter of fact, 18:2n-6 has long been known as being required for water permeability of skin. Arachidonic acid (20:4n-6) has been reported as a messenger fatty acid by itself, acting upon several signaling targets, including protein kinase C. Docosahexaenoic acid (22:6n-3) is a well-known PUFA required for the adequate functioning of the brain and retina. However, it is hard to be sure that those effects are really due to the non-esterified form of the PUFA rather than their ester forms in cell membrane complex lipids, and/or due to their spontaneous oxygenation either by oxygenases or through non-enzyme peroxidation processes. As a matter of fact, docosahexaenoic acid is believed to exert its mains function in the brain and retina as esterified in membrane phospholipids.

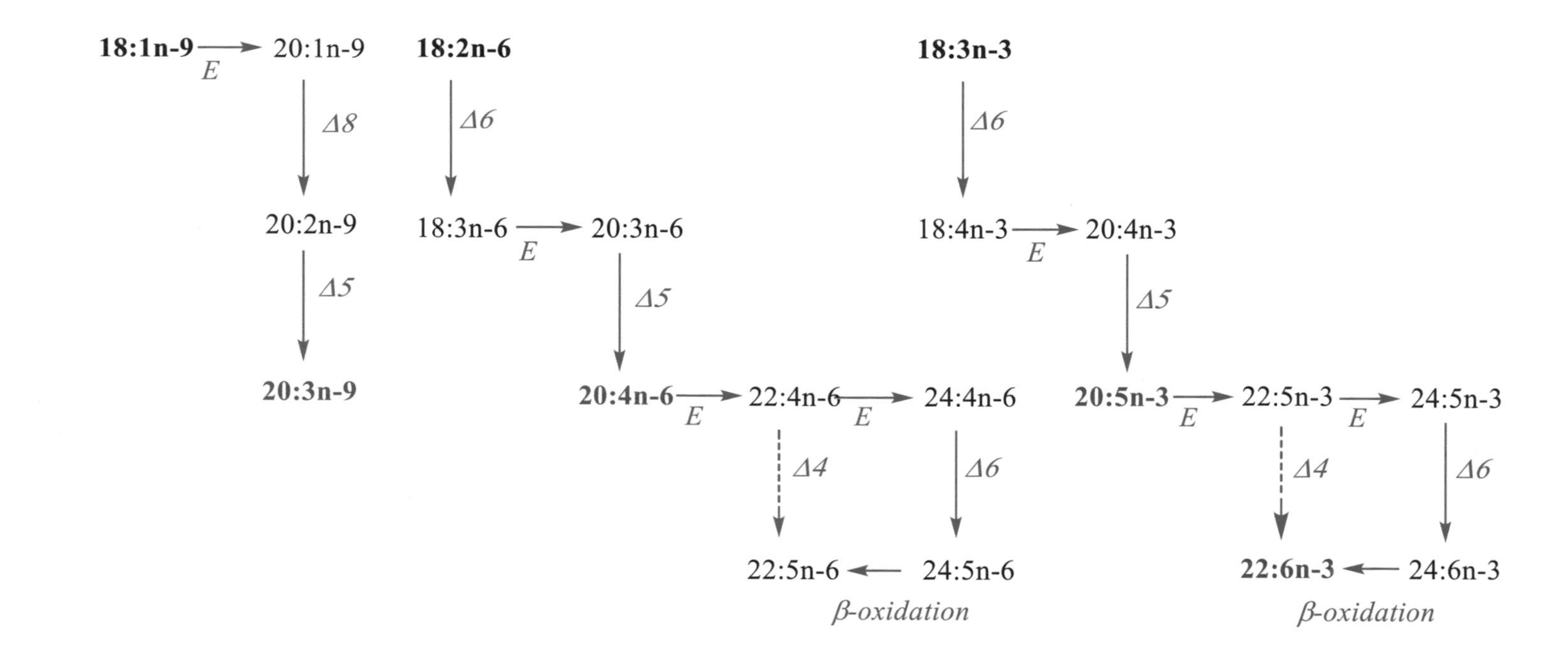

Figure 7 Most abundant polyunsaturated fatty acids (PUFA) in animal cells. The biogenesis of linoleic and alpha-linolenic acids, the essential/indispensable ones, occurs in plants.

Each PUFA is named by the number of carbons followed by the number of methylene-interrupted double bonds, and the position of the double bond closest to terminal or omega carbon. For example, the indispensable precursors are linoleic (18:2n-6) and alpha-linolenic (18:3n-3) acids. Their chemical names are 9*Z*,12*Z*-octadecadienoic and 9*Z*,12*Z*,15*Z*-octadecatrienoic acids, respectively. Z means that the double bond between carbons 9 and 10 or 12 and 13 has the cis/*Z* geometry. Thus, the chemical nomenclature numbers carbons from the carboxylic end instead of the methyl end, although the nomenclature from this end (n- or omega) reflects that elongations and desaturations in animals occur from the constant position of the unsaturated carbons closest to the methyl end towards the carboxylic end.

In case of n-6 PUFA deficiency, the animals may use oleic acid (18:1n-9), that can be made from stearic acid (18:0) by Δ9 desaturase, to make Mead acid (20:3n-9) that partly mimics 20:4n-6 with three common double bonds close to the carboxylic end.

Two common pathways for the biogenesis of polyunsaturated fatty acids (PUFA) in animals, especially mammals, start from the indispensable PUFA. The two pathways are simplified showing the most relevant intermediates. The three major ones are shown in blue; arachidonic acid (20:4n-6/5*Z*,8*Z*,11*Z*,14*Z*-eicosatetraenoic acid) is the most abundant PUFA in the body, except in the brain and retina, in which docosahexaenoic acid (22:6n-3/4*Z*,7*Z*,10*Z*,13*Z*,16*Z*,19*Z*-docosahexaenoic acid) is the major PUFA.

Δ means desaturase creating one *Z* double bond at carbon indicated by the number (six for Δ*6* or 5 for Δ*5*) whereas *E* means elongase that add two carbons to the carboxylic end of each substrate. Thus, the double bond numbers of an elongation product must be incremented by two. As an example, arachidonic acid is elongated into adrenic acid (22:4n-6/7*Z*,10*Z*,13*Z*,16*Z*-docosatetraenoic acid). It must be noted that desaturases and elongases require that each substrate be activated into acyl-coenzyme A. This does not appear in the figure for simplification.

A dotted arrow is associated with Δ*4* because this direct desaturation only occurs in microalgae. In mammals the pathway has proved to be much more indirect with first an additional elongation, then a desaturation by Δ6, and finally shortening by β-*oxidation* occurring within peroxisomes.

In plants (not shown) the two indispensable fatty acids in mammals, 18:2n-6 and 18:3n-3 are produced by other desaturases (not present in animals) from 18:1n-9, namely Δ*12* and Δ*15* desaturases, respectively.

o Prostanoids

Prostanoids belong to a big family of cyclic and oxygenated derivatives from few PUFA (four of them), 20:4n-6 being the reference from which all the characteristic metabolic pathways have been described. Corresponding derivatives from the three others (20:3n-6, 20:5n-3, and 22:4n-6) have been reported by analogy with 20:4n-6 metabolism.

The name prostanoids covers a family of oxygenated products derived from the initial cyclooxygenation of those PUFA. Taking the example of 20:4n-6, the cyclooxygenase enzyme activity of the prostaglandin (PG) H synthase protein, oxygenates and cyclizes first 20:4n-6 into PGG_2 that is further reduced into PGH_2 by the hydroperoxidase activity of the same protein (Fig. 8). It is noteworthy that cyclization consumes two double bonds, meaning that two remaining double bonds are present in products from 20:4n-6. Further from the intermediate PGH_2 that has a short half-life, the opening of the cyclic endoperoxide by PGD, PGE, or PGFα synthase, the latter being a reductase, leads to PGD_2, PGE_2, and $PGF_{2\alpha}$, respectively. The former two PGs can also be produced by physical-chemical degradation of PGH_2, which is facilitated by plasma albumin. Two cytochrome P_{450}-related isomerases called thromboxane and prostacyclin synthases actively convert PGH_2 into thromboxane (Tx) A_2 and prostacyclin or PGI_2, respectively, that are highly unstable derivatives (30 s and 2 min half-lives, respectively, in biological media). TxA_2 and PGI_2 are easily hydrolyzed into the stable and inactive metabolites TxB_2 and 6-oxo-$PGF_{1\alpha}$, respectively.

Two iso-enzymes of PGH synthase have been well characterized: PGHS-1 or COX-1 and PGHS-2 or COX-2. PGHS-1 is known as the constitutive form that is ubiquitously distributed within tissues as a house keeping enzyme. PGHS-2 is the inducible enzyme that is expressed in response to inflammatory stimuli, and under oxidative stress conditions. Both enzymes are membrane-bound homodimers. PGHS-1 is mainly associated with the endoplasmic reticulum and secondarily with the nuclear membrane, whereas the contrary has been reported for PGHS-2. Both enzymes are inhibited by non-steroidal anti-inflammatory drugs (NSAIDs) such as aspirin and indomethacin. Although aspirin, or acetyl salicylic acid, is an "old" drug, it appears to be the most specific inhibitor in acetylating a serine residue, Ser 530 or 529 (depending on the animal species) and Ser 516 of PGHS-1 and -2, respectively, of the active site of the enzyme. This acetylation prevents the substrate PUFA to access the site, and to be properly cyclo-oxygenated. This completely blocks prostanoid production, although some oxygenation without cyclisation may occur with PGHS-2 which has a bigger active site than that of PGHS-1. Subsequently, acetylated PGHS-2 remains capable of producing 11-hydroxy-eicosatetraenoate (11-HETE) from 20:4n-6, as a cyclooxygenase aborted product. As an important application of the different sizes

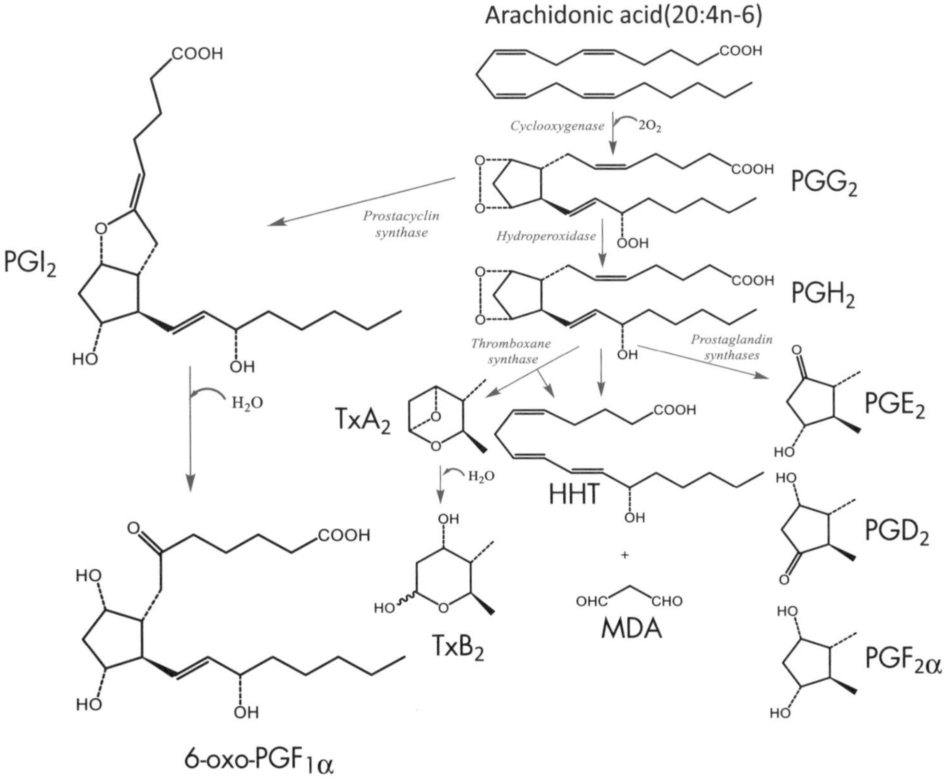

Figure 8 Biosynthesis of prostanoids and related compounds from arachidonic acid (20:4n-6).

This figure gives the example of prostanoid synthesis from arachidonic acid, the reference PUFA for this oxygenated metabolism. They all belong to the series two, meaning that two double bonds remain after oxygenation and cyclization of 20:4n-6, except for 6-oxo-$PGF_{1\alpha}$ because one double bond in PGI_2 disappears when degraded by hydration.

The main structural modifications relate to the first step catalyzed by cyclooxygenase, leading to PGG_2. The second step is catalyzed by the hydroperoxidase activity of prostaglandin (PG) H synthase. The other alterations from PGH_2 are catalyzed by specific prostaglandin synthases or isomerases, including for the non-PG thromboxane (Tx) formation. Another quantitatively important route is the cleavage of PGH_2 into 12-Hydroxy-Heptadeca-Trienoic (HHT) acid plus malondialdehyde (MDA), a water soluble molecule that is often used as a global marker of oxidative stress (see Other peroxidation products, pp. 44-51). Prostacyclin and thromboxane synthases are cytochrome P_{450} isomerases.

of the active site of PGHS enzymes, new NSAIDs able to enter the PGHS-2 active site, but not that of PGHS-1, have been designed to selectively inhibit the inflammatory-induced PGHS-2, and keep active the house keeping PGHS-1 that is usually requested for normal functioning of certain cells. This is well-known in stomach where some prostaglandin formation is required to regulate the proton transport. The abolishment of prostaglandin production may then be responsible for ulcer generation.

It may be noted that PGH_2 can also be spontaneously cleaved at the level of its pentagonal ring-associated cyclic endoperoxide, to release a C_3 entity keeping the two oxygen atoms, called malondialdehyde (MDA), and its C_{17} counterpart called 12-hydroxy-heptadecatrienoate (12-HHT). Such a cleavage can be catalyzed by thromboxane synthase as well, which makes this step a quantitatively efficient one. So far, no relevant bioactivity has been reported for 12-HHT, but it is worth stating that MDA is a by-product of cyclooxygenase activities, then it cannot be considered solely as a marker of non-enzymatic lipid peroxidation, as frequently considered (see Other peroxidation products, pp. 44-51).

Apart from TxB_2 and 6-oxo-$PGF_{1\alpha}$, the other prostanoids all possess potent biological activities. TxA_2 and PGI_2 have a very short half-lives (as already stated above), and exhibit opposite effects, i.e. induction and inhibition of platelet aggregation, vasoconstriction, and vasodilatation, respectively. They act through specific seven trans-membrane receptors coupled with G proteins, Gq (activation of PLC_β) for the former and Gs (activation of adenylyl cyclase) for the latter (Table 1).

Table 1 Summary showing prostanoid receptor types and subtypes, with their coupling to G proteins, and meaning in terms of cell second messengers proteins.

Prostanoid:	TxA_2	PGI_2		PGE_2			$PGF_{2\alpha}$	PGD_2	
Receptor subtype:			(EP1)	(EP2)	(EP3)	(EP4)		(DP1)	(DP2)
Nb of amino-acid residues:	369	386	402	358	390	488	359	366	395
G proteins	Gq	Gs	Gq	Gs	Gi	Gs	Gq	Gs	Gi

DP: D prostaglandin; EP: E prostaglandin.
Gi: hetero-trimer G protein $\alpha_i\beta\gamma$; Gq: hetero-trimer G protein $\alpha_q\beta\gamma$; Gs: hetero-trimer G protein $\alpha_s\beta\gamma$.

The other prostaglandins, PGD_2, E_2 and $F_{2\alpha}$, are called primary prostaglandins because they are known from the 1960s. They have more diverse roles such as pro-inflammatory actions on various target cells, or constricting activity toward the uterine smooth muscle for $PGF_{2\alpha}$. They all act through seven trans-membrane receptors coupled with G proteins (Table 1). PGE_2 is rather peculiar as at least three receptor subtypes, coupled with Gi (inhibition of adenylyl cyclase) or Gs, leading to the opposite activation of adenylyl cyclase, but also

Gq with the activation of phospholipase C_{β}. This multiple G protein coupling is not unique as it has also been described for adrenergic receptors. As a matter of fact, a well-known model is that of blood platelets that bear two receptor subtypes, 2 and 3 (E prostaglandins EP2 and EP3), coupled with Gs and Gi, respectively. The biological response then depends on the affinity of PGE_2 for the two subtypes, with a higher affinity for the EP3 than for the EP2. So, a low concentration of PGE_2 leads to a decreased cyclic AMP (resulting in the activation of platelet aggregation), and a higher concentration does the contrary (with the inhibition of platelet aggregation through an increased cyclic AMP).

In addition to these membrane receptors coupled with hetero-trimeric G proteins, some nuclear receptors have been identified. They are the transcription factors peroxisome proliferator-activating receptors (PPAR). As a matter of fact, we may mention the identification of PPARα as a receptor for LTB_4 (see Leukotrienes, pp. 34-40), PPARβ/δ for prostacyclin, and PPARγ for 15-deoxy-Δ^{12}-PGJ_2 (see below the double dehydration of PGD_2).

As PGD_2, E_2 and $F_{2\alpha}$ are relatively stable molecules, apart from some dehydration of the PGD_2 and E_2 cycle into PGJ_2 and PGA_2/B_2, respectively, the most potent way to abolish their biological activity is the dehydrogenation/oxidation of their secondary alcohol by a rather ubiquitous NAD-dependent 15-PG dehydrogenase into their 15-oxo derivatives. Such a modification, followed by the reduction of the *trans/E* double bond at carbon 13, leads to the completely inactive 15-oxo-13,14-dehydro prostaglandins (Fig. 9A).

In addition, prostanoids being fatty acids, they are concurrently degraded by β-oxidation into dinor- and tetranor-PGs (Fig. 9) that are more specifically excreted into urine where they serve as useful markers for the production of prostanoids in the whole body.

One specific dehydrogenation of TxB_2 into 11-dehydro-TxB_2 has been reported as a prominent pathway, likely because of the facilitated oxidation of the secondary alcohol next to the oxane (Fig. 9B). 11-dehydro-TxB_2 in urine is considered as a reliable marker of TxA_2 production *in situ*. Concurrently, the beta-oxidation process also acts on the stable thromboxane TxB_2 (Fig. 9B).

A special mention must be made relating to the double dehydration product of PGD_2. As stated above, the PGD_2 cycle may be dehydrated into PGJ_2 that has been described as a mitogenic product, and further isomerized and dehydrated on the methyl chain to provide 15-deoxy-Δ^{12}-PGJ_2 (15dPGJ_2) (Fig. 10). Much attention has been paid to this metabolite since it has been shown to specifically activate the peroxisome proliferator-activated receptor (PPAR) gamma (see above). As such, several biological activities have been reported in response to 15dPGJ_2. In contrast to PGD_2 dehydration products, the PGE_2 dehydration derivatives PGA_2 (produced in acidic conditions) and PGB_2 (in basic/alkaline conditions) (Fig. 10) are generally considered as inactive products, although some studies have reported biological activities associated with PGA_2.

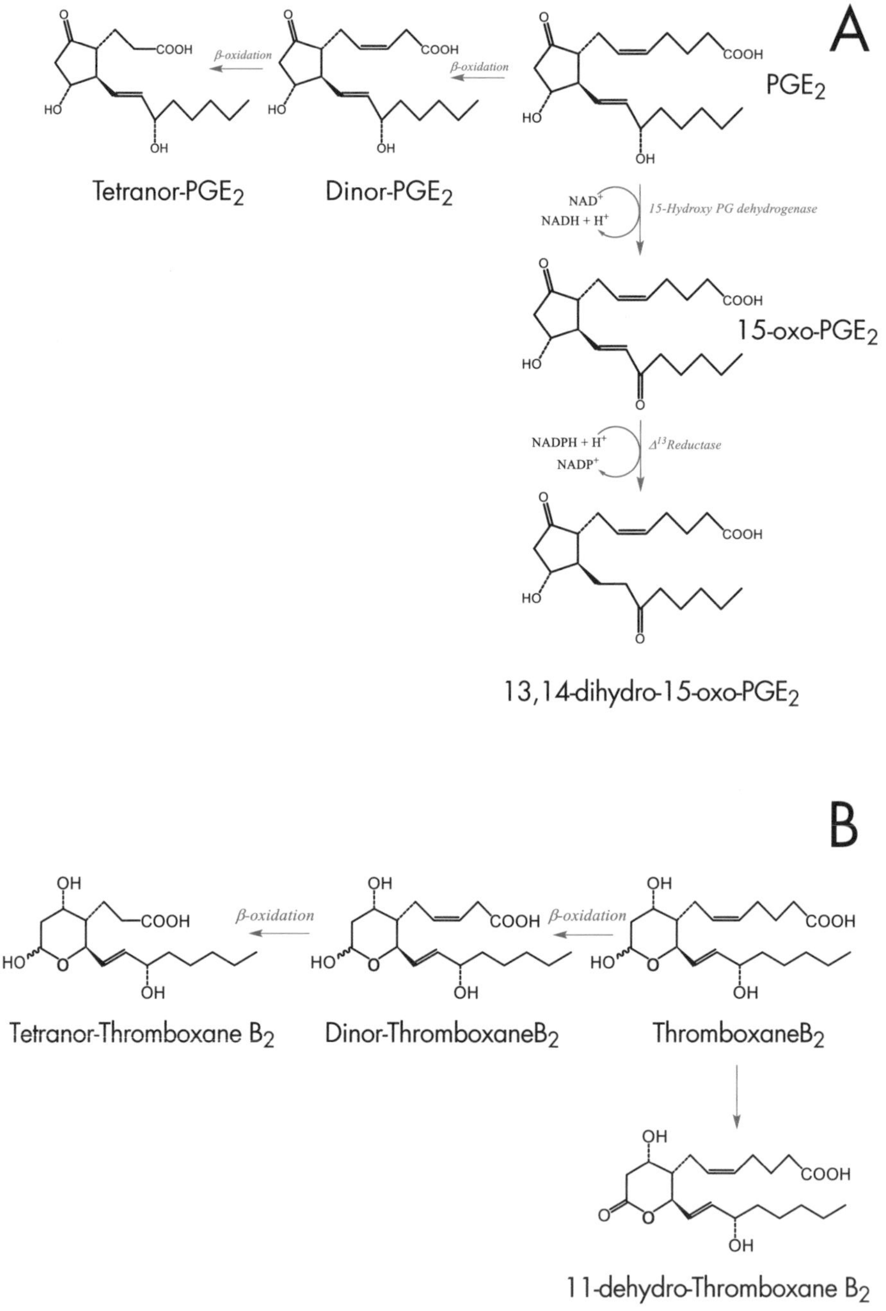

A

Tetranor-PGE_2 ← β-oxidation ← Dinor-PGE_2 ← β-oxidation ← PGE_2

NAD^+ / $NADH + H^+$ — 15-Hydroxy PG dehydrogenase

15-oxo-PGE_2

$NADPH + H^+$ / $NADP^+$ — Δ^{13}Reductase

13,14-dihydro-15-oxo-PGE_2

B

Tetranor-Thromboxane B_2 ← β-oxidation ← Dinor-ThromboxaneB_2 ← β-oxidation ← ThromboxaneB_2

11-dehydro-Thromboxane B_2

Figure 9 Main pathways for the degradation of prostanoids.

The common and efficient route to degrade prostanoids is first the dehydrogenation of the secondary alcohol at carbon 15, followed by reduction of the double bond next to the resulting ketone. The figure shows the example of PGE_2 (Fig. 9A). In addition, TxB_2 is dehydrogenated at carbon 11 that is presumably facilitated by the alpha-oxygen of the oxane cycle. 11-oxo-TxB_2, usually called 11-dehydro-TxB_2, is considered as a reliable urinary marker of TxB_2 from blood origin (Fig. 9B).

Another common degradation route is β-oxidation that occurs like with classical fatty acids, i.e. on prostanoyl-CoA. Such a β-oxidation is likely to be limited to tetra-nor derivatives (two cycles of β-oxidation) because of the proximal cyclopentane or oxane (Fig. 9A & 9B).

Much less data are available for prostanoids derived from 20:3n-6, 20:5n-3, and 22:4n-6. They can, however, be summarized as follows:

- 20:3n-6 (dihomo-gamma-linolenic acid/DGLA) is converted into the series one prostanoids (PG_{1S}: 5,6-dihydro-PG_{2s}) compared with the series two from 20:4n-6, at the exception of the prostacyclin counterpart PGI_1 that cannot be synthesized because of lack of double bond at carbon 5. Of note, PGH_1 is produced at similar rate as PGH_2 from the same concentrations of precursor PUFA, but much less TxA_1/B_1 is produced compared to TxA_2/B_2, which means that thromboxane synthase is relatively specific of PGH_2. Within the series one, only PGE_1 (Fig. 11) has been found of interest as an activator of adenylyl cyclase in target cells. Its receptor has not been well characterized, but this prostaglandin is supposed to act through the prostacyclin receptor.

- 20:5n-3 (eicosapentaenoic acid/EPA), as a major PUFA of marine lipids, reported of great interest in the prevention of cardiovascular risk, has been more thoroughly investigated. Among the series three prostanoid products (PG_{3S}: Δ17-PG_{2s}) (Fig. 11), it can be stressed that PGI_3 is as potent as PGI_2, whereas TxA_3 is almost devoid of aggregation and vasoconstricting activities compared to TxA_2, and is produced in lower amount compared to TxA_2, confirming the rather high specificity of thromboxane synthase for PGH_2. Altogether, these specificities give an interesting mechanism of action for the biological activity of their precursor 20:5n-3, as it is considered as a protective agent against cardiovascular events through generation of high prostacyclin potential but low thromboxane activity. Also, PGE_3 seems to be less pro-inflammatory than PGE_2, and PGD_3 is an antagonist of PGD_2 when the latter promotes endothelium permeability. PGD_3 then appears as an anti-inflammatory agent in contrast to PGD_2. PGD_3 is also dehydrated to PGJ_3 and further converted to 15d-PGJ_3 which has been shown to increase adiponectin secretion by adipocytes. Adiponectin being known to act against atherosclerosis associated with obesity and diabetes, this is a contribution of PGD_3 to protection against cardiovascular diseases. Overall, prostanoids from 20:5n-3

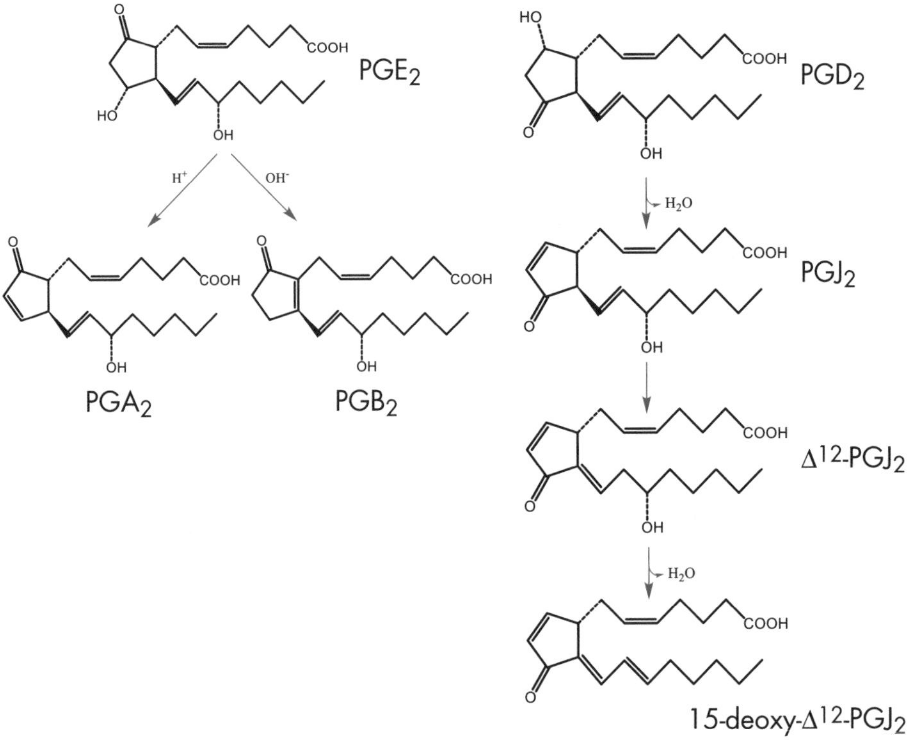

Figure 10 Dehydration of PGD_2 and PGE_2 into cyclopentenones.

PGE_2 may spontaneously be dehydrated with creation of a double bond in the cyclopentane ring due to the presence of the oxo group. This is limited at neutral pH but facilitated at acidic pH (leading to PGA_2) or at basic pH (leading to PGB_2). PGA_2 can be converted into PGB_2 and *vice-versa* through the intermediate PGC_2 (with an intermediate double bond in the cyclopentane ring compared to PGA_2 and PGB_2). PGA_2 has been reported with some biological activity, but this is controversial. In contrast, PGB_2 is considered inactive. It has been used as an internal standard in HPLC using UV detection (at λmax: 270 nm) for the measurements of leukotrienes (see Fig. 14) because of its conjugated triene including the oxo group on the cyclopentene.

PGD_2, a position isomer of PGA_2, is preferentially dehydrated into PGJ_2, further isomerized into Δ12-PGJ_2, which allows a second dehydration in the methyl chain. This sequential dehydration process is of high biological relevance since both products (PGJ_2 and Δ12-PGJ_2) are bioactive molecules with different activities (see Prostanoids, pp. 24-31), whereas the catabolism stated in Figure 9 relates to loss of biological activities of the products.

can be taken as active products against athero-thrombogenesis. Also, considering that thromboxane synthase is rather specific for PGH_2, the result will be a higher formation of prostaglandins of the series three, such as PGD_3 (as well as of the series one), by diversion of the metabolic routes, which is a way to contribute in the anti-atherothrombotic potential of 20:5n-3.

- 22:4n-6 (adrenic acid) is converted into di-homologs of PG_{2s} (Fig. 11), but very few functional data are available on them. It is generally assumed that dihomo-prostanoids are much less active than prostanoid counterparts from 20:4n-6, as especially shown with dihomo-PGI_2 and dihomo-TxA_2 compared to PGI_2 and TxA_2.

The degradation products of PG_{1s}, PG_{3s} and dihomo-PG_{2s} have not been substantially investigated. It is assumed that they undergo the same kind of inactivation through 15-PGDH and β-oxidation.

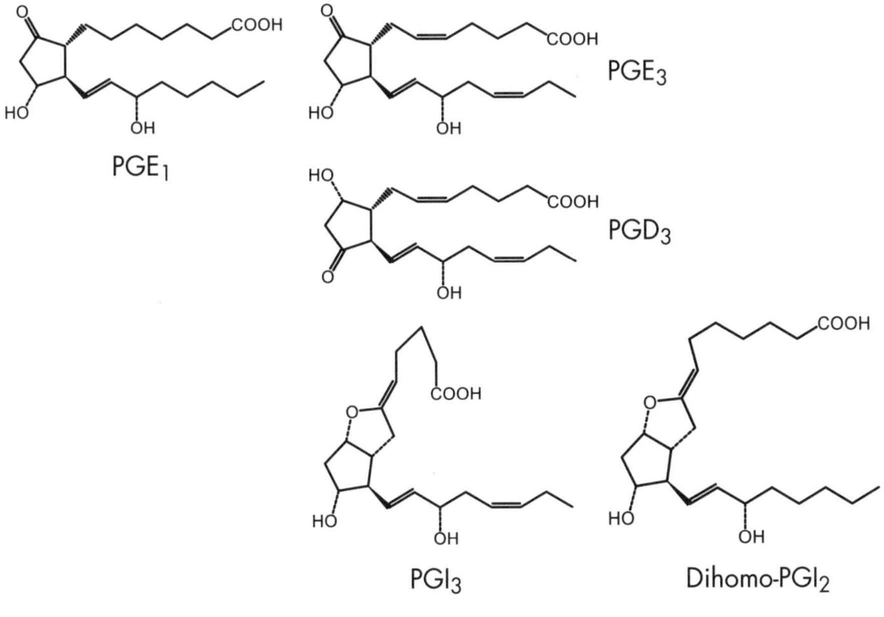

PGE1 PGE3 PGD3 PGI3 Dihomo-PGI2

Figure 11 Prostanoids of biological interest from 20:3n-6, 20:5n-3 and 22:4n-6.

Five prostanoids produced from those three PUFA have clearly been described for their biological interest. PGE_1, of the series one (isomer of PGE_2 with only one double bond, as the precursor 20:3n-6 has three double bonds instead of four in arachidonic acid). PGD_3, E_3 and I_3, of the series three issued from 20:5n-3, isomers of PGD_2, E_2 and I_2, respectively. The fifth prostanoids is the dihomolog dihomo-PGI_2, because it derives from adrenic acid (22:4n-6). Its first double bond is at position 7 instead of 5 in PGI_2.

o Lipoxygenase products

By definition, lipoxygenases are dioxygenases of lipids, first described in plants, where the abundant 18:2n-6 appears as a fairly good substrate of the n-6-lipoxygenases. In principle, lipoxygenases oxygenate all PUFA, so their fatty acid specificities are much broader than that of PGH synthases. Their common mechanism is to abstract a hydrogen atom/radical from the methylene group of a 1,4-cis,cis-pentadiene motif in PUFA, which leads to the double bond conjugation with a *cis,trans* isomerization due to some stabilization of the initial radical:

The isomerization shown in this figure occurs towards R_1 but may occur towards R_2 as well, depending on the specificity of the lipoxygenase involved.

The new radical binds one molecule of oxygen, leading to a peroxyl radical, which becomes a hydroperoxide by adding the hydrogen radical initially abstracted and transiently stored by the enzyme. The enantio-specificity of the hydrogen abstraction usually leads to *S* derivatives due to the asymmetric carbon oxygenated, but few lipoxygenases make *R* derivatives. Such a hydroperoxide is usually reduced into the corresponding hydroxyl derivative by glutathione peroxidases (GPx), mainly cytosolic if the hydroperoxide substrate is a non-esterified PUFA, or the so-called phospholipid hydroperoxide GPx if the substrate hydroperoxide is in complex lipids, especially glycerophospholipids (Fig. 12). GPx are selenoprotein enzymes that have one atom of selenium in a seleno-cysteine residue in their active site. This seleno-cysteine is directly involved in the reduction process converting the hydroperoxide into the hydroxyl. As a matter of fact, mutated GPx with a cysteine residue in place of the seleno-cysteine, considerably decreases the reduction rate.

Lipoxygenases require the minimal structure called 1,4-cis/Z,cis/Z-pentadiene (in red color within the substrate). They abstract one hydrogen atom from the methylene group, which is pro-chiral in the presence of the enzyme, called anterofacial hydrogen. The resulting radical in the fatty chain is stabilized by migration of one of the adjacent double bond (specific of a given lipoxygenase), which leads to another radical, with the shifted cis/*Z* double bond that becomes trans/*E*. Thus, a cis,trans/*Z,E* conjugated diene is generated.

A dioxygen is then attached to the latter radical to form a peroxyl radical which is stabilized by the addition of the initial hydrogen atom kept by the enzyme. The position of the dioxygen attached to the radical is opposite to that of the initial abstracted hydrogen atom, considering the substrate in a plan. The oxygenated carbon is usually of the (S) configuration.

The resulting fatty hydroperoxide may be reduced into its corresponding secondary alcohol by glutathione peroxidase (GPx), usually the most abundant isoenzyme GPx-1. The latter reduction can be summarized as follows: ROOH + 2GSH → ROH + GS-SG + H_2O. (GSH: reduced glutathione; GS-SG: oxidized glutathione).

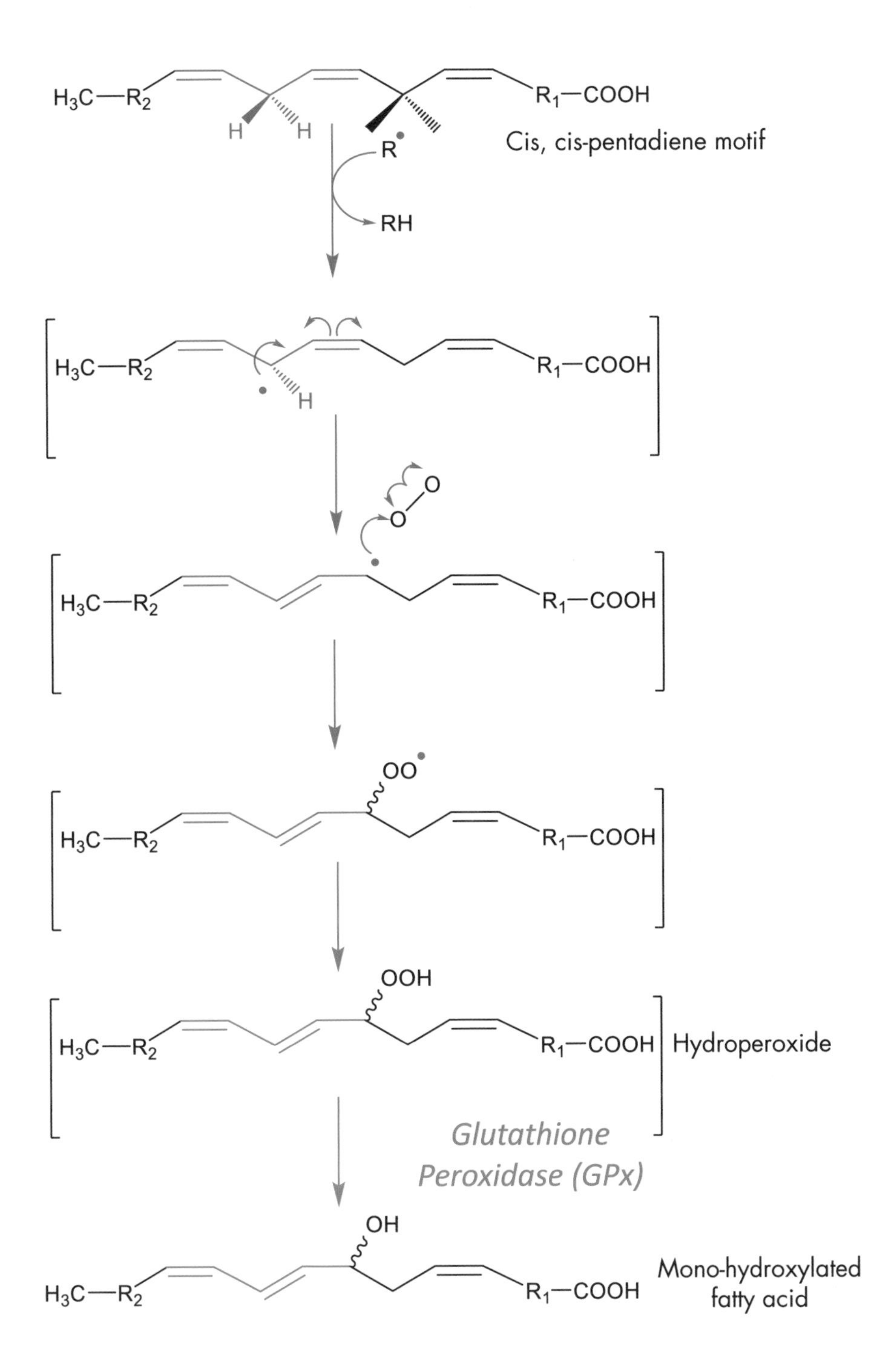

Figure 12 General mechanism for lipoxygenation of PUFAs and further reduction of the product by glutathione peroxidases.

Thus, a great number of oxygenated derivatives can be produced from PUFA, because of various sites of oxygenation in the carbon chains and the number of PUFA. In addition, the highest unsaturated PUFA may undergo several oxygenations of the same molecule, namely double or even triple lipoxygenation. The end-products after reduction of the hydroperoxide intermediates are mono-, di- or tri-hydroxylated derivatives.

From ArA, the reference molecule, a series of HETEs (hydroxy-eicosatetraenoates) such as 5-, 8-, 12-, and 15-HETE have been described through the so-called 5-, 8-, 12-, and 15-lipoxygenases, and further reduction by GPx (Fig. 13). Also, 8(S),15(S)-diHETE has been reported as a double lipoxygenation product. As the methylene moiety for the initial abstraction of hydrogen is selected from the methyl terminus by 15-lipoxygenase, the enzyme may also be called n-6-lipoxygenase, because it may produce 13- and 17-hydroperoxides from C18 and C22 PUFA, respectively, in addition to 15-hydroperoxide from C20 PUFA such as 20:4n-6. This is also valid for 12-lipoxygenase that acts as an n-9-lipoxygenase.

Several animal lipoxygenases (LOX) have been described according to the carbon number for the oxygenation of arachidonic acid. The main LOX are often called 5-, 12- and 15-LOX. Since the 12- and 15-LOX recognize the substrate from the methyl terminus, they may also be called n-9 and n-6 -LOX, allowing the application to other substrates at different carbons from their methyl terminus. As an example, 15-/n-6-LOX will convert 18:2n-6 into 13-hydroperoxy-ocyadecadienoic acid (13-HpODE) and 22:6n-3 into 17-hydroperoxy-docosahexaenoic acid (17-HpDoHE).

The products from ArA are hydroperoxy-eicosa-tetraenoic acids (HpETE) that are further reduced by GPx-1 (see Fig. 12) into hydroxyl-eicosa-tetraenoic acids (HETE).

– *Leukotrienes*

Leukotrienes have been named according to their structural characteristic, that is a double conjugation resulting in a conjugated triene, and because they were first discovered in blood leukocytes.

The first description of this sort was done with the enzyme called 5-lipoxygenase because it specifically oxygenates carbon 5 of 20:4n-6. Contrary to other lipoxygenases, 5-lipoxygenase further makes a 5,6-epoxide derivative which generates the conjugated triene geometry (7E,9E,11Z) in the first leukotriene (LT) of the series, called LTA_4. This leukotriene is unstable because of the epoxide located in the alpha position of the conjugated triene. It can easily be hydrolyzed into LTB_4 (5S,12R-dihydroxy-6Z,8E,10E,14Z-eicosatetraenoic acid) by LTA_4 hydrolase (Fig. 14), but it is unstable enough to be

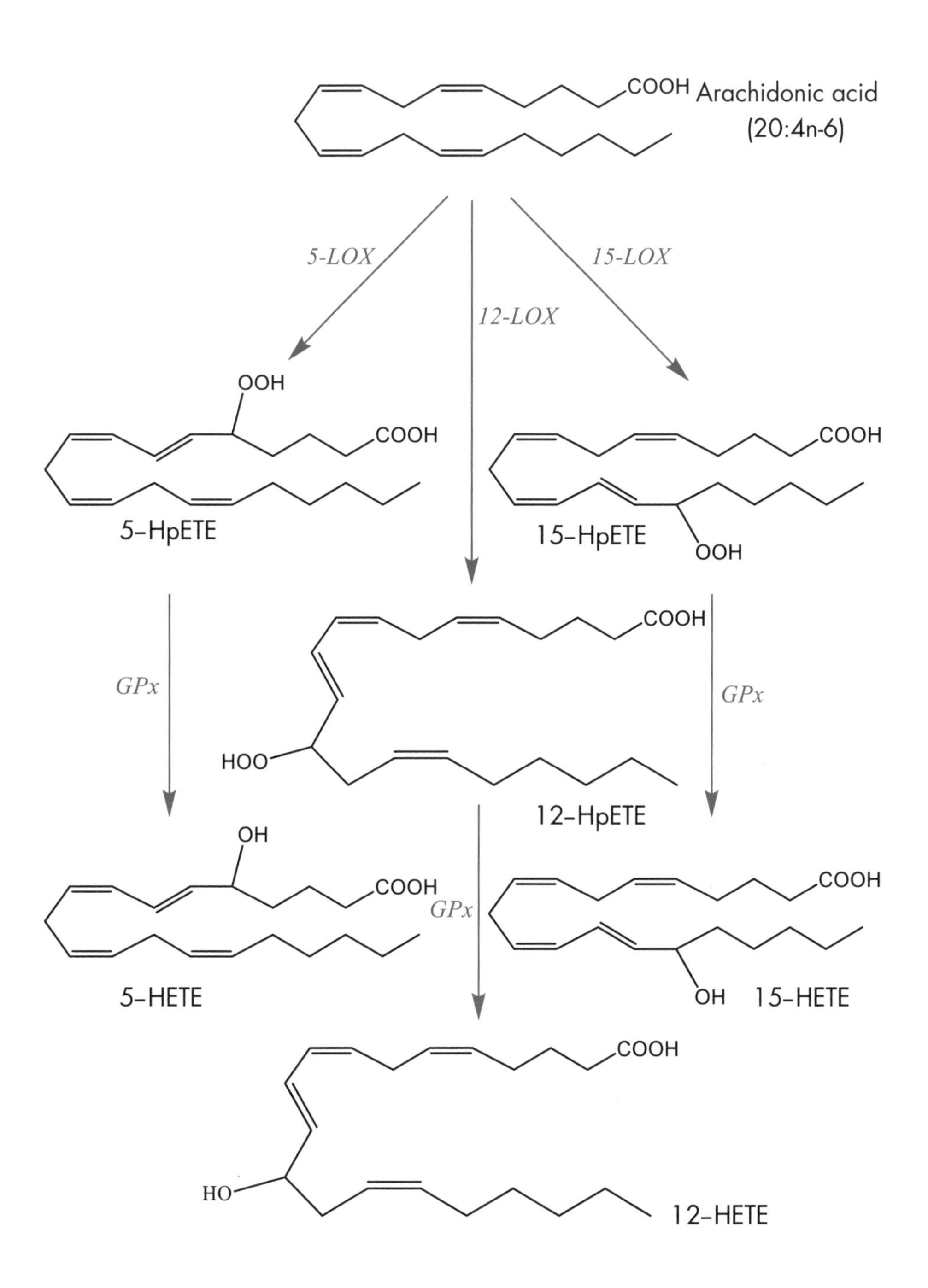

Figure 13 Main hydroxy-eicosatetraenoates (HETEs) of interest produced by lipoxygenation of ArA.

spontaneously hydrolyzed into two LTB_4 isomers (12-epi-LTB_4 & all-*trans* conjugated triene LTB_4). LTB_4 is a potent chemotactic agent and thus exhibits a pro-inflammatory profile. As for the double lipoxygenation of 20:4n-6 by 15-lipoxygenase into 8(S),15(S)-diHETE, an isomer of LTB_4, called LTBX, is the end-product of 5- and 12-lipoxygenases acting successively. LTBX is 5S,12S-dihydroxy-6E,8Z,10E,14Z-eicosatetraenoic acid.

The intermediate LTA_4 can also be conjugated with reduced glutathione (GSH) by glutathione-S-transferase to generate the first peptido-leukotriene of the series called LTC_4. This conjugation results in the epoxide opening to produce 5S-hydroxy,6-glutathionyl-7E,9E,11Z,14Z-eicosatetraenoic acid (LTC_4). Then, LTC_4 can undergo the removal of the γ-glutamate moiety from the GSH residue by γ-glutamyl transferase to produce a glutamate derivative and the second peptide-leukotriene LTD_4. The mixture LTC_4/LTD_4 exhibits potent broncho-constricting activities as well as vaso-constricting ones. LTD_4 may be further degraded into LTE_4 by a peptidase that removes the glycine residue (Fig. 14). It is assumed that LTE_4 is a partially inactive product within the various degradation products that follow.

LTB_4 production is predominant in neutrophil polymorphonuclear leukocytes whereas peptide-LTs predominate in eosinophils.

In addition to what is shown for the lipoxygenation of ArA by 12- and 15-LOX, the first product of 5-LOX, 5-hydroperoxy-eicosa-tetraenoic acid (5-HpETE), may be converted by the same enzyme into 5,6-epoxy-eicosatetraenoic acid, called leukotriene A_4 (LTA_4), while part of it is also reduced into 5-hydroxy-eicosatetraenoic acid (5-HETE) by glutathione peroxidase (see Figure 13). The epoxide formation from the conjugated diene HpETE induces a further conjugation. The geometry of the resulting conjugated triene is *E,E,Z*.

The product, LTA_4, is quite unstable because of the double bond next to the epoxide. The presence of LTA_4 hydrolase efficiently converts LTA_4 into LTB_4 but, because of LTA_4 is unstable, a substantial part may be spontaneously hydrolyzed into LTB_4 isomers (not shown). LTB_4 results from the addition of a hydroxyl group at carbon 12 with the R stereochemistry, and the opening of the epoxide. LTB_4 is 5(S),12(R)-di-OH-6*Z*,8*E*,10*E*,14*Z*-eicosatetraenoic acid. Therefore the *7E,9E,11Z* geometry of the conjugated triene in LTA_4 is shifted to the *6Z,8E,10E* conjugated triene in LTB_4.

Another route converts LTA_4 into peptido-LTs by addition of reduced glutathione (GSH), as catalyzed by glutathione S-transferase. The immediate product is LTC_4, obtained by opening the 5,6-epoxide with GSH keeping the S configuration of the secondary alcohol at carbon 5, the S-glutathionyl being R. LTC_4 is further transformed into LTD_4 by γ-glutamyl transferase which transfers the glutamate moiety to another substrate. LTD_4 is further transformed into LTE_4 by a specific peptidase which releases glycine. The production of LTE_4 is often considered as a substantial loss of peptido-LTs activity. In all peptido-LTs the initial 7*E*,9*E*,11*Z* geometry of the conjugated triene in LTA_4 is kept the same.

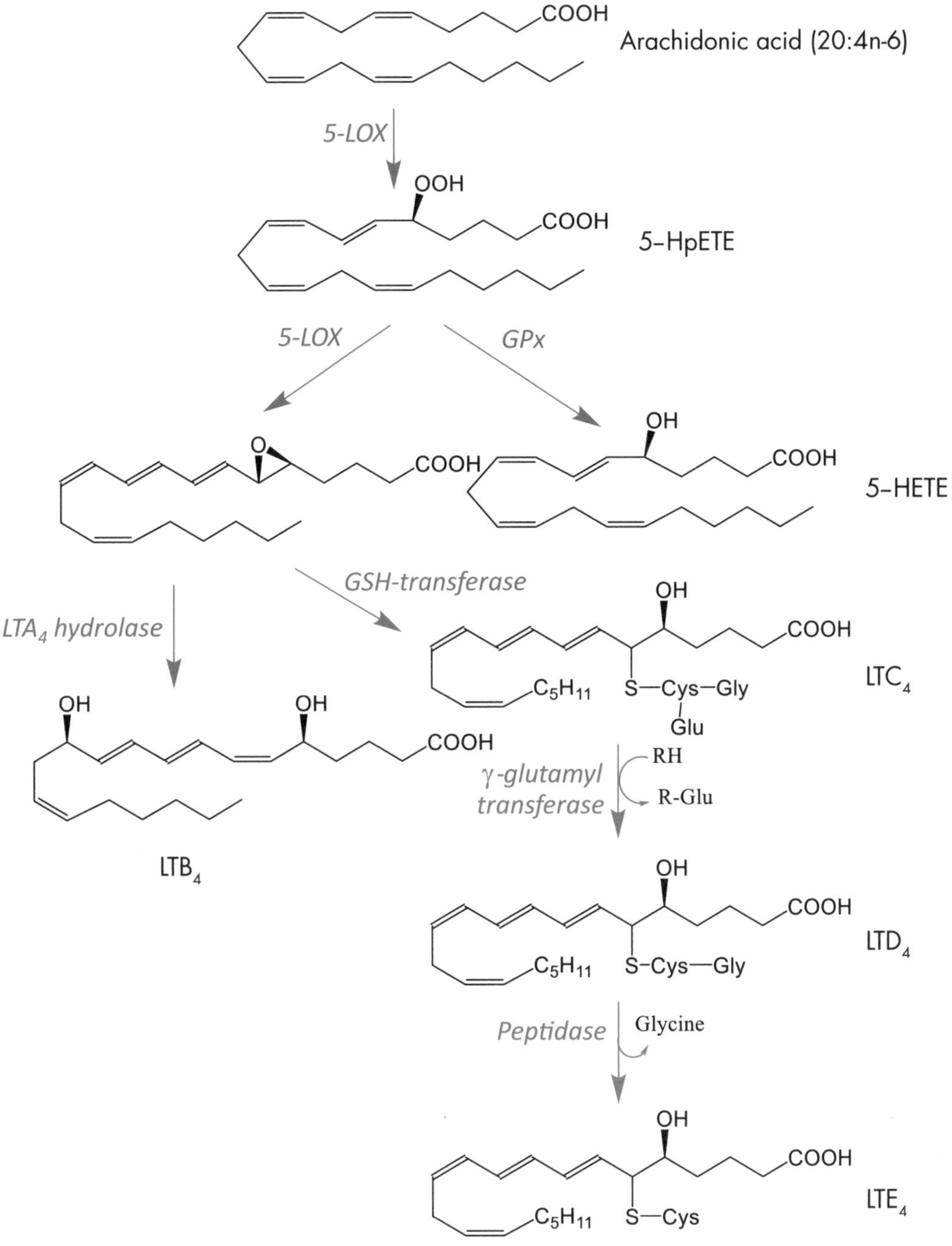
COOH
Arachidonic acid (20:4n-6)
5-LOX
OOH
COOH
5–HpETE
5-LOX
GPx
O
COOH
OH
COOH
5–HETE
GSH-transferase
LTA4 hydrolase
OH
COOH
C5H11
S—Cys—Gly
Glu
LTC4
OH
OH
COOH
LTB4
γ-glutamyl transferase
RH
R-Glu
OH
COOH
C5H11
S-Cys—Gly
LTD4
Peptidase
Glycine
OH
COOH
C5H11
S—Cys
LTE4

Figure 14 Formation of leukotrienes from ArA.

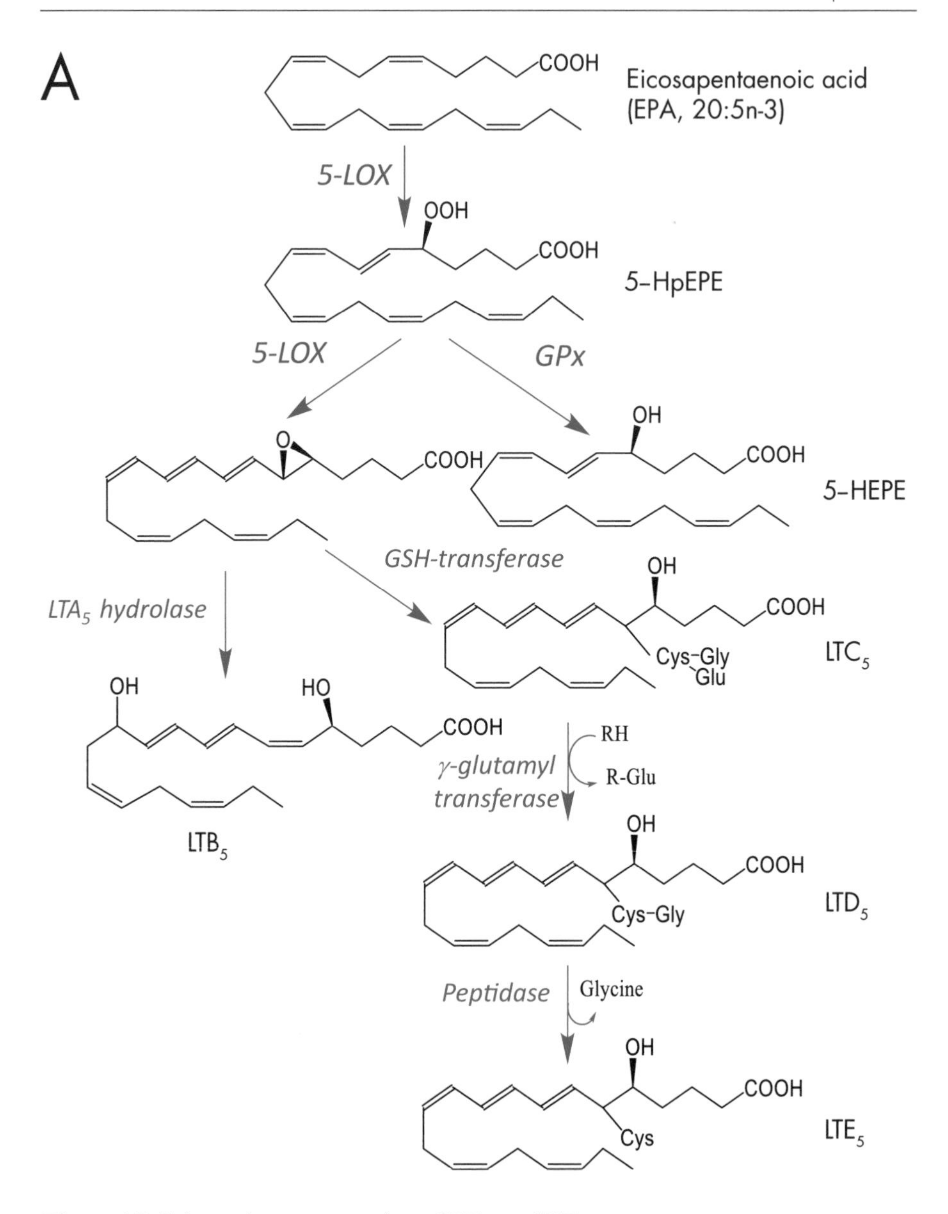

Figure 15 Schematic representation of LT_{3S} and LT_{5s}.

Production of leukotrienes from eicosapentaenoic acid (20:5n-3), namely LT_{5s}, is assumed to follow the same route as for LT_{4s} (Fig. 15A). The stereochemistry of the corresponding LTs is also assumed to be the same in both series. The only difference in the former is the additional 17*Z* double bond present in the initial precursor 20:5n-3.

Figure 15B shows the same scheme for LT_{3s} production from Mead acid (20:3n-9), the n-6 PUFA deficiency marker. The only difference from LT_{4s} is of course the lack of double bond out of the conjugated triene as the 14*Z* one is missing in the precursor 20:3n-9.

B

COOH
Mead acid (20:3n-9)
5-LOX
OOH
COOH
5-HpETrE
5-LOX
GPx
O
COOH
OH
COOH
5-HETrE
GSH-transferase
LTA$_3$ hydrolase
OH
COOH
C_6H_{11}
S—Cys—Gly
Glu
LTC$_3$
OH
OH
COOH
γ-glutamyl transferase
RH
R-Glu
LTB$_3$
OH
COOH
C_6H_{11}
S-Cys—Gly
LTD$_3$
Peptidase
Glycine
OH
COOH
C_6H_{11}
S—Cys
LTE$_3$

Taking into account the structural pre-requisite to make leukotrienes according to those from ArA, only two other PUFA can be substrates for such a biosynthesis. They are EPA and 20:3n-9/Mead acid, the marker of the n-6 PUFA deficiency. The products are LTB$_5$/C$_5$/D$_5$, or delta-17 derivatives of LTB$_4$/C$_4$/D$_4$, respectively, and LTB$_3$/LTC$_3$/D$_3$, or dihydro-14,15 derivatives of LTB$_4$/C$_4$/D$_4$, respectively (Fig. 15A & B). Very few is known about the biological activities of LT$_{3s}$, whereas LT$_{5s}$ have been studied in this respect. Roughly, LTB$_5$ is considered at least ten-fold less chemotactic than LTB$_4$, which is in agreement with advices for consuming marine fat that contains EPA to reduce the inflammatory status. LTC$_5$ and LTD$_5$ also have slightly less/similar broncho- and vaso-constricting activities than LTC$_4$/D$_4$, which cannot contribute to the anti-atherothrombogenic potential of EPA.

PDX (10(S),17(S)-diOH-4*Z*,7*Z*,11*E*,13*Z*,15*E*,19*Z*-docosahexaenoic acid) is produced through a double oxygenation by 15-/n-6-LOX. The initial hydroperoxides at position 17 and 10 (likely 17 and then 10) are reduced into hydroxyl derivatives, presumably by glutathione peroxidase.

PD1 (10(R),17(S)-diOH-4*Z*,7*Z*,11*E*,13*E*,15*Z*,19*Z*-docosahexaenoic acid) is assumed to involve first 15-/n-6-LOX to produce 17(S)-OOH-DHA, followed by formation of the 16,17-epoxide, and hydrolysis as it has been shown for LTB_4 production from ArA. This mechanism, first described in LTB_4 synthesis, requires that 15-/n-6-LOX exhibits, as 5-LOX for LTB_4 does, a second enzymatic activity making the epoxide from the 17 hydroperoxide. Otherwise, it must be supposed that the epoxidation from the hydroperoxide spontaneously occurs.

Maresin R1 (MaR 1) is assumed to be produced the same way as PD1 but with 12-/n-9-LOX of macrophages. Its structure is 7(R),14(S)-diOH-4*Z*,8*E*,10*E*,12*Z*,16*Z*,19*Z*-docosahexaenoic acid.

Contrary to other lipoxygenases, 5-lipoxygenase is a calcium-dependent enzyme to favor its translocation toward the inner leaflet of the plasma membrane, where it binds an activating cofactor protein called FLAP (five-lipoxygenase activating protein).

– *Other hydroxylated products*

Beyond the classic leukotrienes, other oxygenated products of PUFA have been described as double lipoxygenation products having a *trans,cis,trans*/*E,Z,E* conjugated triene in between the two secondary alcohols. This common 1(S),8(S)-dihydroxy-*E,Z,E*-octatriene motif (Fig. 16) has been called “poxytrin” and found to inhibit cyclooxygenases and the aggregatory activity of thromboxane A_2. This geometry motif seems to be crucial for the latter biological activity, since geometric isomers such as protectin D1 or maresin 1 issued from docosahexaenoic acid (DHA), exhibiting a 1(R),8(S)-dihydroxy-*E,E,Z*-octatriene motif (Fig. 16), exert anti-inflammatory effects but no anti-aggregatory activities. These structure-function relationships are especially remarkable for protectin D1 versus its stereo and geometric isomer protectin DX, a member of the “poxytrin” family.

As already mentioned in the previous paragraph, an isomer of LTB_4 may be produced by double lipoxygenation of ArA in using successively 5-LOX and 12-LOX or the other way. This isomer, sometimes called LTBX, (5(S),12(S)-diOH-6*E*,8*Z*,10*E*,14*Z*-eicosatetraenoic acid) is a poxytrin. The poxytrin family also includes 8(S),15(S)-diHETE already mentioned, and two double lipoxygenase products of alpha-linolenic acid (ALA, 18:3n-3): 9(S/R),16(S)-diOH-10*E*,12*Z*,14*E*-octadecatrienoic acid, called linotrins.

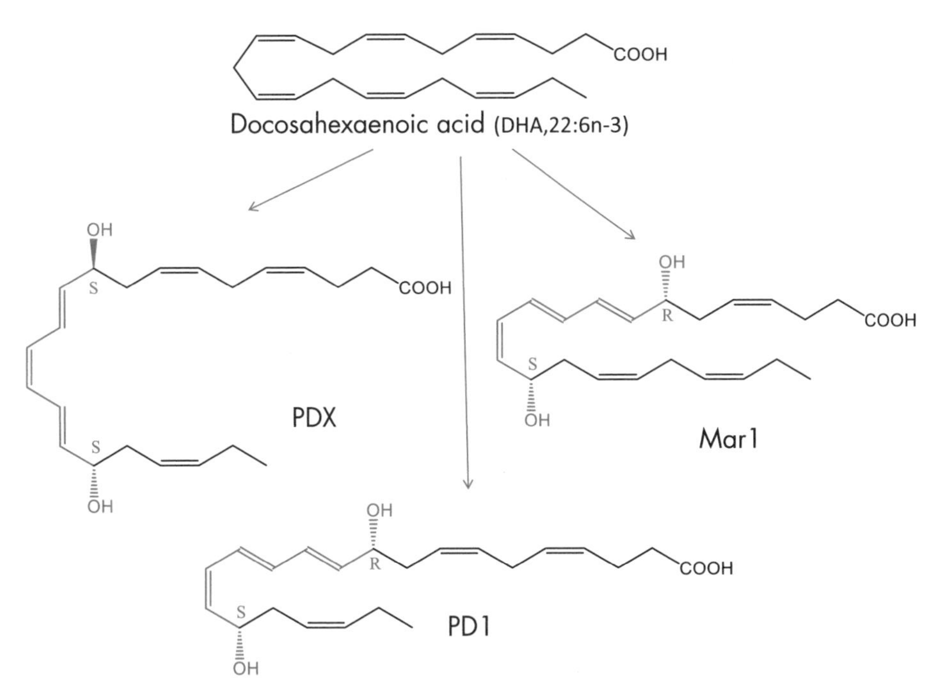

Figure 16 Oxygenated trienes produced by double oxygenation of DHA.

Some PUFA such as EPA and DHA may undergo more oxygenation as they are highly unsaturated. Indeed, anti-inflammatory derivatives called resolvins, because they accelerate the resolution of inflammation, have been described. In most cases they are tri-hydroxylated products with conjugated dienes, trienes and even tetraenes, alone or in combination within the same molecule. They are resolvins E1, E2, E3 from EPA and resolvins D1, D2 and D3 from DHA (Fig. 17).

o Cytochrome P_{450} products

Another way to oxygenate fatty acids and produce bioactive metabolites is the monooxygenase pathway using cytochrome P_{450} as a cofactor. This produces either monohydroxylated derivatives or epoxides that are further hydrolyzed into dihydroxylated products by epoxide hydrolases.

Regarding PUFA to be converted into monohydroxylated products, the omega and omega-1 hydroxylases have been most frequently reported (Fig. 18A). The most well-known products derive from arachidonic acid; e.g. 20-HETE has been described as a vasoactive product.

In addition, the omega hydroxylation is assumed to be one step in the degradation of other oxylipins. As a matter of fact, LTB_4 is first converted into

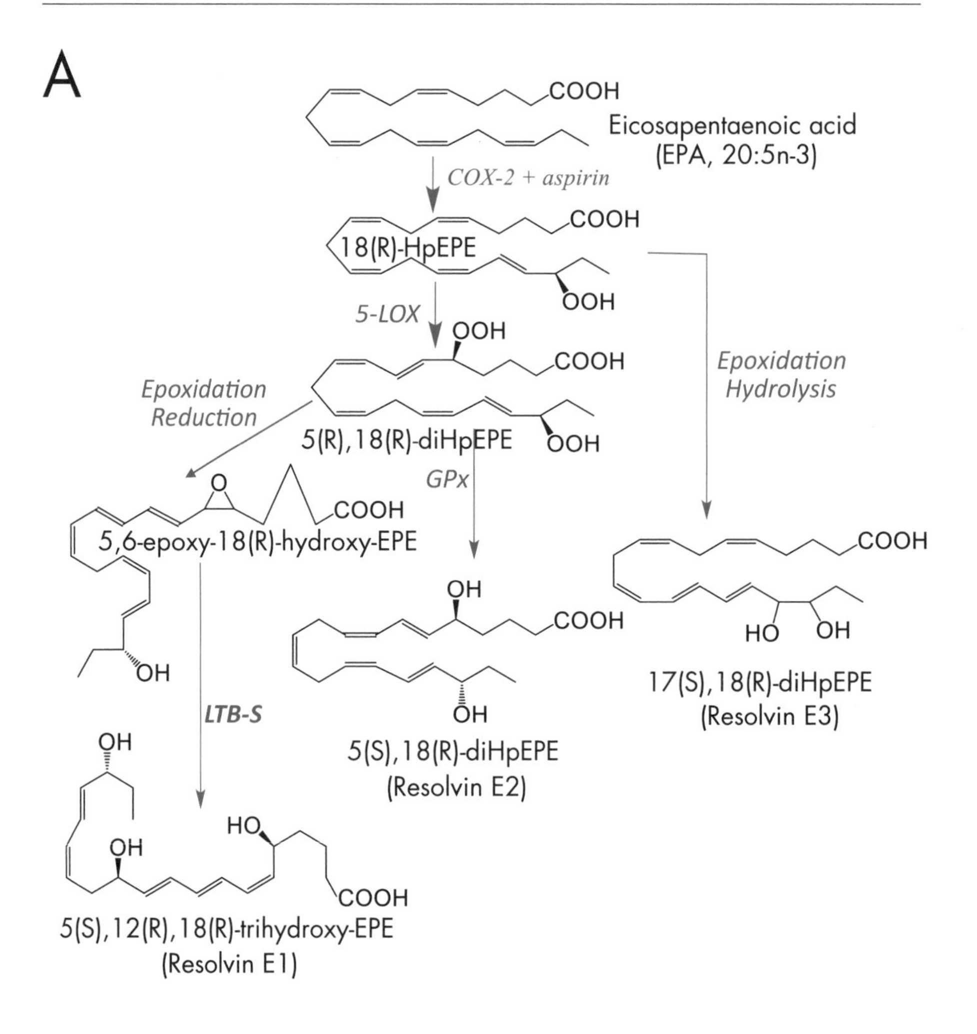

Figure 17 Resolvins from EPA and DHA.

Resolvin production from EPA and DHA are complex processes. Figure 17 gives some examples in two specific situations starting from the oxygenation of EPA by aspirinated COX-2 (Fig. 17A) and the 15-LOX oxygenation of DHA (Fig. 17B).

When COX-2 is acetylated by aspirin treatment, EPA may be oxygenated at carbon 18 instead of being cyclooxygenated into PGG_3 in absence of aspirin. The product is 18(R)-hydroperoxy-eicosapentaenoic acid (18(R)-HpEPE), that can be further oxygenated by 5-LOX into 5(S),18(R)-diHpEPE, which undergoes the classical effect of 5-LOX, i.e. the 5,6 epoxidation followed by the hydrolysis by the LTA hydrolase or LTB synthase (LTB-S), and the reduction of the 18-OOH, resulting in 5(S),12(R),18(R)-triHEPE named resolvin E1, which is 18(R)-OH-LTB_5.

Alternatively, the first 5-Lox product 5(S),18(R)-diHpETE may be reduced by GPx into 5(S),18(R)-dHEPE named resolvin E2, or the initial 18(R)-HpEPE undergoes epoxidation followed by hydrolysis into 17(S),18(R)-diHEPE (resolvin E3).

B

COOH
Docosahexaenoic acid (DHA, 22:6n-3)
15-LOX
HOO COOH HO
7(S)-Hydroperoxy-
17(S)-hydroxy-DHA
HOO COOH HO
4(S)- Hydroperoxy-
17(S)-hydroxy-DHA
COOH HOO
17(S)-HpDHA
O COOH HO
7(S),8(S)-epoxy-
17(S)-hydroxy-DHA
O COOH HO
4(S),5(S)-epoxy-
17(S)- hydroxy-DHA
HO OH COOH HO
Resolvin D1
HO COOH OH HO
Resolvin D2
OH HO COOH HO
Resolvin D3

Figure 17B shows the production of resolvins from DHA. The first product in the process is the 15-lipoxygenation of DHA into 17-OOH-DHA or 17-HpDoHE, then reduced into 17-HDoHE which may be further lipoxygenated by 5-LOX into 7-OOH,17OH-DHA or 7-OOH,17HDoHE. The latter undergoes the 5-LOX-induced 7,8-epoxidation followed by a LTA hydrolase-like hydrolysis, leading to 7(S),16(R),17(S)-triHDoHE named resolvin D2. Alternatively, the epoxide intermediate can be hydrolyzed to make 7(S),8(R),17(S)-triHDoHE named resolvin D1.

A third alternative is described in which 17(S)-HpDoHE is converted by 5-LOX into its 4(S)-hydroperoxide derivative, followed by the classical process of epoxidation and its hydrolysis through the LTA hydrolyse-like activity. The final product is the conjugated triene and conjugated diene 4(S),11(R),18(S)-trihDoHE named resolvin D3.

Resolvins D1 and D2, as well as their epoxide intermediates are conjugated tetraenes instead of trienes as in ArA- and EPA-derived leukotrienes.

20-hydroxy-LTB_4, accompanied with an important loss of chemotactic activity that is completely abolished by further oxidation into the 20-carboxylic product of LTB_4. The first conversion into 20-OH-LTB_4 is cytochrome P_{450}-dependent, whereas the additional oxidation is done by NAD^+-dependent dehydrogenases (Fig. 18A). This omega-oxidation of many other oxygenated metabolites from PUFA is likely, but remains to be properly described.

A series of cytochrome P_{450}-dependent epoxygenases have been described acting upon ArA to make 5,6-, 8,9-, 11,12-, and 14,15-epoxy derivatives by epoxidation at the expense of each double bond initially present at those positions. The products are called epoxy-eicosatrienoates (EET), e.g. 5,6-EET. Those epoxides are relatively stable, compared to TxA_2, in which one epoxide is located next to another one, and LTA_4 in which the epoxide is located next to the conjugated double bond. They are, however, rapidly hydrolyzed by epoxide hydrolases into the corresponding vicinal dihydroxylated derivatives 5,6-, 8,9-, 11,12-, 14,15-dihydroxy-eicosatrienoates (diHETrEs) (Fig. 18B). All EETs exhibit biological activities, especially vaso-activities, whereas the diHETrEs are generally devoid of bio-activities. Also, some epoxygenase metabolites of EPA have been described with biological activities.

Figure 18A shows the omega/omega-1 hydroxylation of arachidonic acid (20:4n-6). This is catalyzed by cytochrome (Cyt) P_{450}-dependent monooxygenases, using NADPH as a co-factor. These so-called omega/omega-1 hydroxylases have many other fatty acid substrates. Of interest in functional lipids, LTB_4, the prominent 5-lipoxygenate metabolite of ArA, is inactivated by an omega hydroxylase into 20-OH-LTB_4, and then into 20-COOH-LTB_4 by further dehydrogenation via a double NAD^+-dependent dehydrogenase. This omega oxidation is considered as the inactivation of LTB_4.

Figure 18B describes another specific oxygenation of ARA through Cyt-P_{450}, resulting in epoxides made each at the expense of one double bond. Four different epoxides have been described as eicosatrienoic acids (EETs). Each of them is easily converted by epoxide hydrolases leading to diOH-eicosatrienoic acids (diHETrEs), in most cases inactive compared to the precursor EETs that are vasoactives.

- Other peroxidation products

All the oxidation processes described in the three previous paragraphs (Prostanoids pp. 24-31 to Cytochrome P_{450} products pp. 41-44) deal with enzyme-dependent oxygenation, by either di- or mono-oxygenases. In addition to that, a non-enzymatic oxygenation may occur and is more commonly called lipid peroxidation, similarly to that from dioxygenases making initial lipid peroxides. It is, however, assumed that such a non-enzyme dependent peroxidation depends on various reactive oxygen species (ROS).

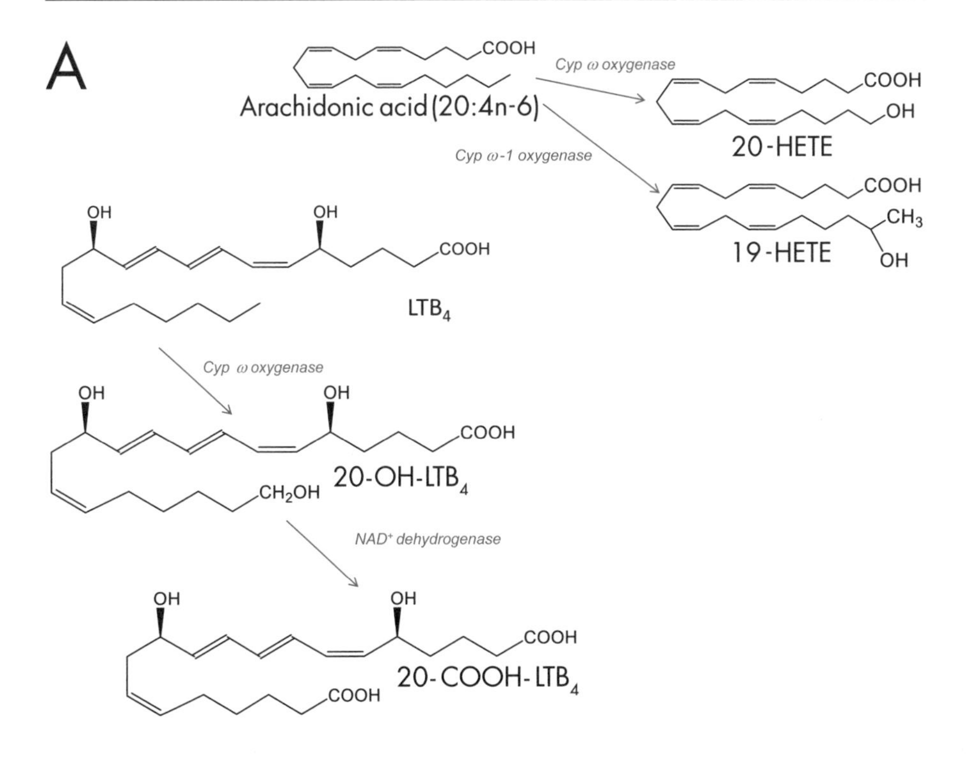

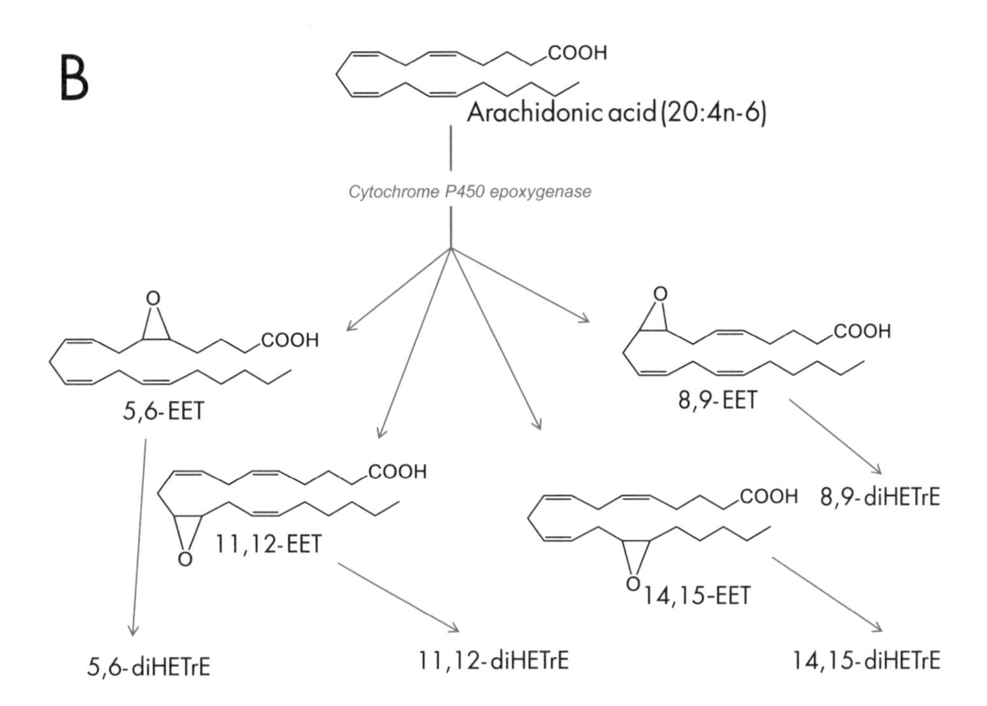

Figure 18 Omega oxidation of ArA and LTB_4; epoxygenation of ArA and hydrolysis.

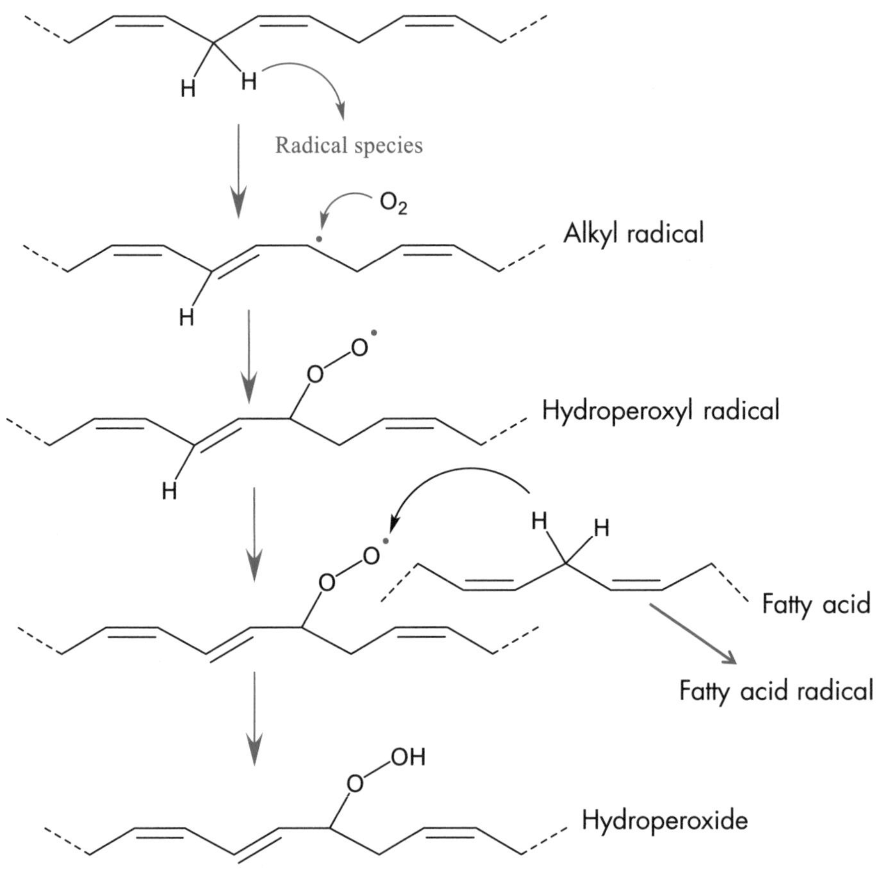

Radical species

O_2

Alkyl radical

Hydroperoxyl radical

Fatty acid

Fatty acid radical

Hydroperoxide

Figure 19 General scheme for non-enzyme dependent lipid peroxidation.

The 1,4-cis/Z,cis/Z-pentadiene structure loses a hydrogen radical abstracted by another radical (frequently an oxygen radical). The resulting organic radical is quickly "stabilized" by a double bond shift to make a conjugated diene (the shifted double bond becomes *E* that is more stable than the *Z* form). Then the newly formed organic radical reacts with non-radical dioxygen to make a peroxyl radical, easily stabilized into the corresponding hydroperoxide by taking one hydrogen radical from another molecule, thus generating a new radical (propagation process). In contrast to what is stated in Figure 12 for the lipoxygenase mechanism, either *Z* double bond may be shifted, allowing two position isomers in the peroxyl radical formation. (Figure 19 shows only the *E/Z* conjugation from left to right, but the opposite *Z/E* conjugation also occur with equal frequency in absence of lipoxygenase).

This scheme is very similar to lipoxygenase-induced hydroperoxide formation (Fig. 12), except that the product is racemic (R/S) whereas it is usually purely (S) via lipoxygenation.

The non-enzyme-dependent mechanism of lipid peroxidation is similar to that described with lipoxygenases (Fig. 12), with a hydrogen abstraction from the methylene located in between two non-conjugated double bonds of a non-esterified or esterified PUFA, by a radical, which could be either an oxygen radical or an organic radical. The resulting radical is quickly isomerized into a *Z,E* conjugated diene radical that binds a dioxygen molecule to make a peroxyl radical, which can abstract a hydrogen radical from a surrounding PUFA to be transformed into a hydroperoxide PUFA (Fig. 19). This cascade process is called the propagation phase of lipid peroxidation. It can occur on non-esterified PUFA and in PUFA esterified in complex lipids such as glycerolipids and sterylesters.

Biological systems are equipped with glutathione peroxidases (GSH-Px) to reduce hydroperoxides into the corresponding hydroxylated non-esterified or esterified PUFA. Two main GSH-Px are involved: GSH-Px-1 acts upon non-esterified hydroperoxides, and GSH-Px-4, called phospholipid GSH-Px to state that phospholipid hydroperoxides are their preferred substrates, although it may also reduce hydroperoxides in cholesteryl-esters (see Lipoxygenase products, pp. 32-40).

Non-esterified as well as esterified hydroperoxides can spontaneously degrade by cleavage into hydroxyl-alkenals. A possible mechanism of this degradation is shown in Figure 20A with the example of 4-hydroxy-2*E*-nonenal (4-HNE) formation. The production of 4-HNE and 4-hydroxy-2*E*-hexenal (4-HHE) preferentially represents n-6 and n-3 PUFA peroxidation, respectively (Fig. 20B). This is expected if the preferred oxygenation of PUFA is proximal to the methyl end. Obviously, the frequency of the cleavage is highest when GSH-Px activities are weak. The presence of a reactive carbonyl (aldehyde) conjugated with a double bond (carbons 1 & 2) makes hydroxy-alkenals able to easily condense with primary amines, producing Schiff bases, or making Michael adducts on the *trans* double bond. This is especially relevant to protein modification of their lysyl and cysteinyl residues, respectively. Hydroxyl-alkenals may be further metabolized by oxidation of both the carbonyl and hydroxyl groups. As a matter of fact, 4-HNE is further oxidized into 4-hydroxy-nonenoic acid (4-HNA) and/or dehydrogenated into 4-oxo-2*E*-nonenal (4-ONE), the latter being even more reactive to make molecular adducts, due to the presence of two carbonyls separated by one double bond. This is, of course, valid for other hydroxyl-alkenals, such as 4-HHE, although this has been much less documented.

A classic and commonly measured peroxidation end-product is 8-epi-$PGF_{2\alpha}$ or 8-iso-$PGF_{2\alpha}$, a stereoisomer of $PGF_{2\alpha}$ with the carboxylic chain in β-position, versus the cyclopentane ring, as the methyl terminal chain. The two chains are then *cis/Z* relating to the ring whereas they are *trans/E* in all cyclooxygenase-derived prostanoids (Fig. 8). However, 8-epi-$PGF_{2\alpha}$ is only one of the 32 isoprostanes that are end-products of the ROS-dependent arachidonic acid peroxidation.

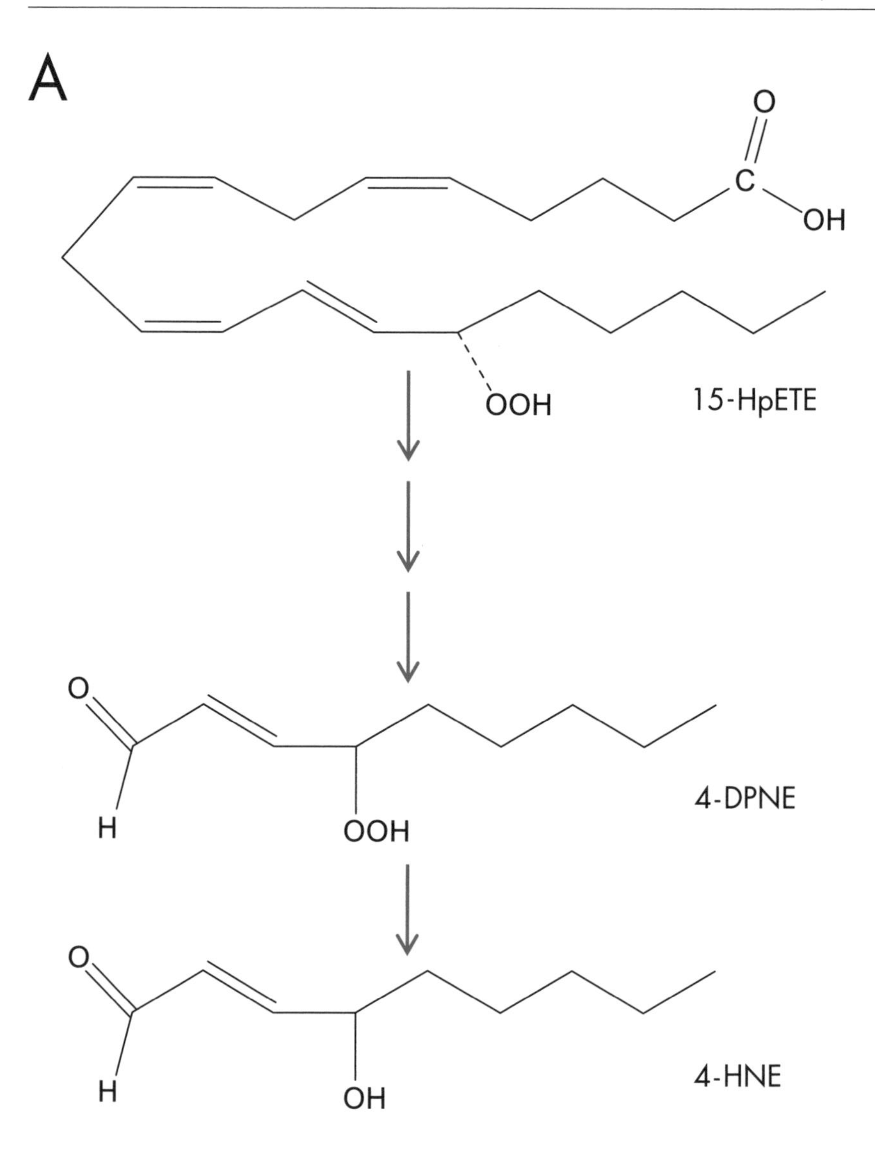

Figure 20 Formation of two types of hydroxyl-alkenals by cleavage of PUFA hydroperoxides.

In Figure 20A, one possible mechanism in the formation of hydroxyl-alkenal is shown. The 15-LOX metabolite from arachidonic acid, 15-HpETE is taken as an example. A second peroxidation occurs in γ-position from the first hydroperoxide, toward the carboxylic end, then a chain cleavage (said Hock cleavage) generates two aldehydes. The carboxylic aldehyde is rapidly oxidized into a dicarboxylic product (not shown) while the second aldehyde keeps its hydroperoxide group. The latter compound, named 4-hydroperoxy-alkenal, is reduced by GSH peroxidase

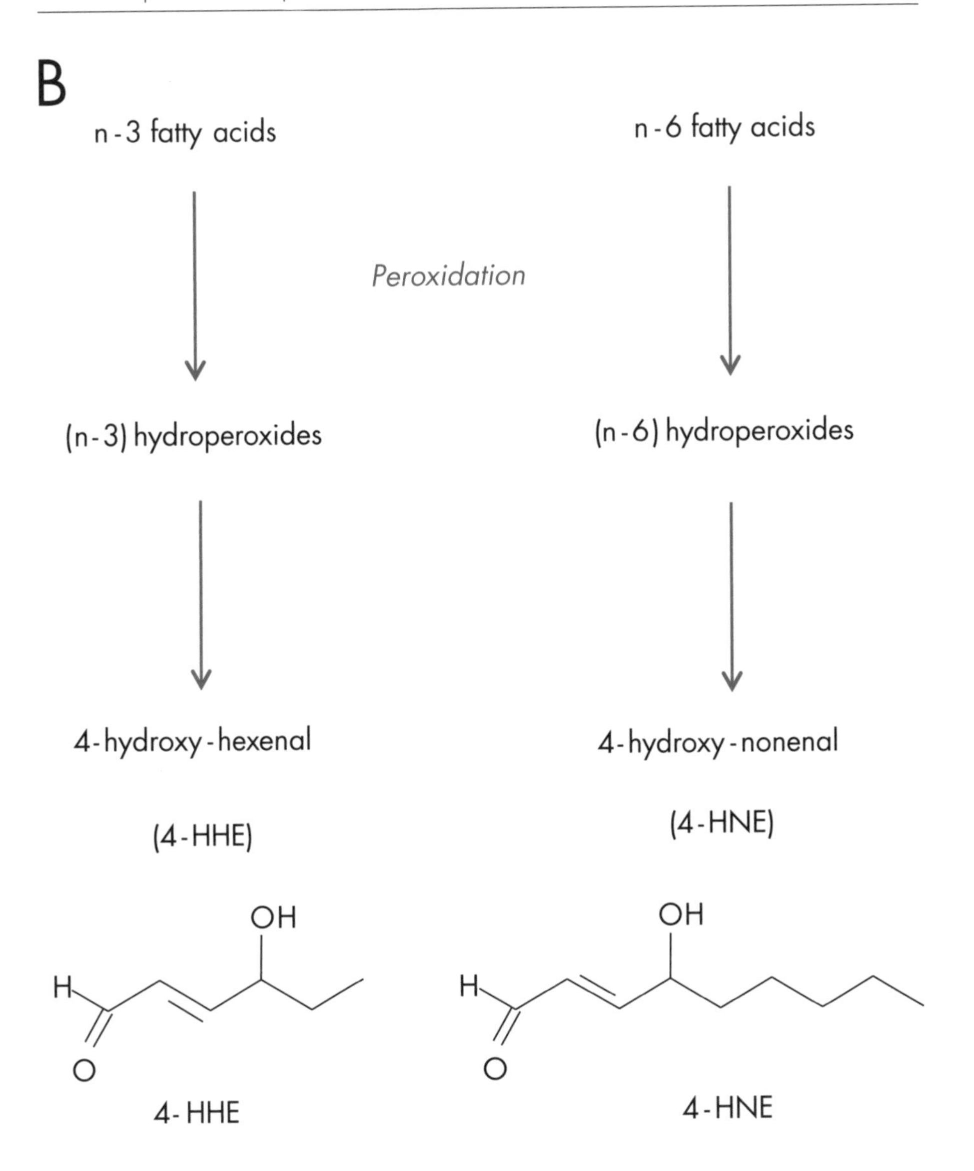

into 4-hydroxy-nonenal (4-HNE), the main lipid peroxidation product from omega-6 PUFA.

Figure 20B shows the formation of the main hydroxyl-alkenals from the most frequent hydroperoxides from omega-3 and omega-6 PUFA. Those hydroperoxides are made distal from the carboxylic end, then providing 4-hydroxy-hexenal (4-HHE) and 4-HNE, respectively.

Because it is a rather prominent one, and has the capacity of contracting the vascular smooth muscle cells via the thromboxane receptor sites, it is usually taken as a marker of lipid peroxidation, although it specifically derives from arachidonic acid, and does not represent the non-enzymatic peroxidation of other prominent PUFA such as linoleic acid and all the n-3 PUFA. The 8-epi-$PGF_{2\alpha}$ counter peroxidation end-product of docosahexaenoic acid is called neuroprostane, and could be named 10-epi-dihomo-$PGF_{4\alpha}$ by analogy with 8-epi-$PGF_{2\alpha}$. No confirmed biological activity has been attributed to this metabolite yet.

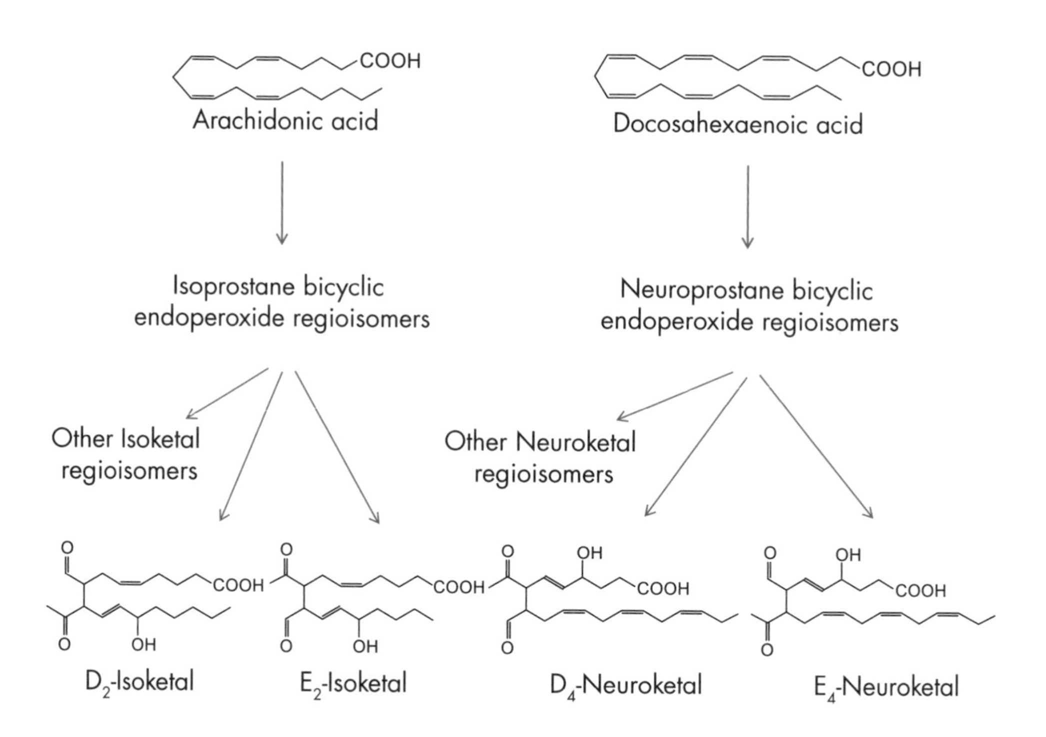

Figure 21 Pathway for the formation of Isoketals (IsoKs) and Neuroketals (NKs) by non-enzymatic free radical-induced peroxidation of ArA and DHA, respectively.

Four/eight IsoP/NP bicyclic endoperoxide regioisomers are formed which then undergo re-arrangement to form four/eight E_2-IsoK/E_4-NK and four/eight D_2-IsoK/D_4-NK regio-isomers. Each E_2-IsoK/E_4-NK and D_2-IsoK/D_4-NK regio-isomer theoretically comprises four/eight racemic diastereo-isomers so that a total of 64/256 regio- and stereo-isomers of IsoKs/NKs can arise from arachidonic/docosahexaenoic acid oxidation, respectively.

Other very reactive products of non-enzymatic lipid peroxidation are iso-ketals from 20:4n-6 and neuroketals from 22:6n-3 (Fig. 21). They mimic the structures of isoprostanes and neuroprostanes but their opened cyclopentane rings exhibit γ-keto-aldehyde moieties that are responsible for covalent binding to primary amines.

A very simple and common derivative issued from lipid peroxidation is malondialdehyde (MDA). It is not so reactive as hydroxyl-alkenals and ketals, but can make bridges between proteins, for instance, due to its two aldehyde groups. This marker is easy to measure but can hardly be considered as reliable to evaluate non-enzymatic lipid peroxidation because it may be produced by thromboxane synthase as well (see also Prostanoids, pp. 24-31 and Fig. 8), and may even derive from the oxidation of sugars. For all these reasons, MDA may not be considered as a specific marker of lipid peroxidation, but the best marker of overall oxidation processes when it is properly and specifically measured.

2.1.2.4 Oxidized phospholipids

Non-enzymatic oxidation of membrane phospholipids generates a very large number of molecules named oxidized phospholipids. The oxidized phospholipids generated depend on the fatty acid present at the *sn*-2 position, the phospholipid subclass, and the oxidant species. Major substrates for oxidation are PUFA, mainly linoleic acid and arachidonic acid, esterified at the *sn*-2 position of glycero-phospholipids. Several well-characterized oxidized phospholipids are stable products derived from the oxidation of 1-palmitoyl-2-arachidonoyl-sn-glycero-3-phosphocholine (PAPC), a prominent molecular species of phosphatidylcholine (PC) in animals. Oxidative cleavage of PAPC generates two molecular species: 1-palmitoyl-2-(5-oxovaleroyl)-sn-glycero-3-phosphocholine (POVPC) and 1-palmitoyl-2-glutaroyl-sn-glycero-3-phosphocholine (PGPC). In addition, an oxidized and cyclized derivative of the ararchidonoyl moiety in PAPC has been reported and named PEIPC (Fig. 22). Many derivatives/fragmentation products have been identified but structural identification of oxidized phospholipids is still incomplete. In particular, molecular species derived from the degradation of 1-palmitoyl-2-linoleoyl-sn-glycero-3-phosphocholine, the main molecular species in blood plasma, remain to be fully characterized, although the radical 13-HODE (13-OH-18:2) replacing linoleoyl at the sn-2 position seems to be such a reliable peroxidation marker.

Plasma levels of oxidized phospholipids are associated with increased risk of coronary artery diseases and apoptosis. POVPC and PGPC have been shown to accumulate in minimally oxidized LDL and to be present in atherosclerotic lesions. They are found in apoptotic cells and may contribute to inflammation.

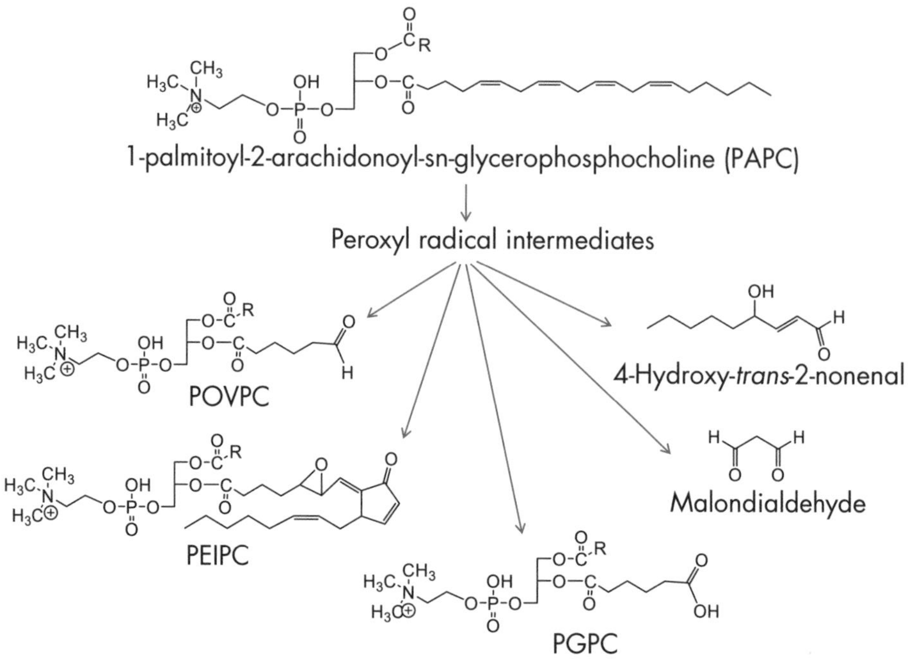

1-palmitoyl-2-arachidonoyl-sn-glycerophosphocholine (PAPC)

Peroxyl radical intermediates

POVPC

4-Hydroxy-*trans*-2-nonenal

Malondialdehyde

PEIPC

PGPC

Figure 22 Degradation products of phospholipid hydroperoxides derived from palmitoyl, arachidonoyl-glycerophosphocholine.

Free-radical-induced oxidation of PUFA esterified within phospholipids, like 1-palmitoyl-2-arachidonoyl-sn-glycero-3-phosphocholine (PAPC), by hydrogen abstraction from the methylene groups of PUFA produces peroxyl radicals which undergo many consecutive reactions including oxidative fragmentation reactions generating short-chain aldehydes (such as 4-hydroxy-trans-2-nonenal and malondialdehyde) and truncated phospholipids. The latter include species containing terminal aldehydic groups (e.g. 1-palmitoyl-2-(5-oxovaleroyl)-sn-glycero-3-phosphocholine (POVPC)) or carboxylic acid groups at the sn-2 position (e.g. 1-palmitoyl-2-glutaroyl-sn-glycero-3-phosphocholine (PGPC)).

Another possible product is not a cleaved one, but is cyclized and doubly oxidized (PEIPC).

2.1.3 Galacto-lipids

Galacto-lipids are included in a wide class of glycolipids in which the galactose residue is a common component. Within this class of glycolipids, glycosphingolipids are of great biological relevance in nervous tissues (see Glycosylceramides, pp. 63-64).

Some galacto-lipids are more simple molecules than phospholipids, as galactose residue(s) are the polar head of glycero-lipids. They are especially

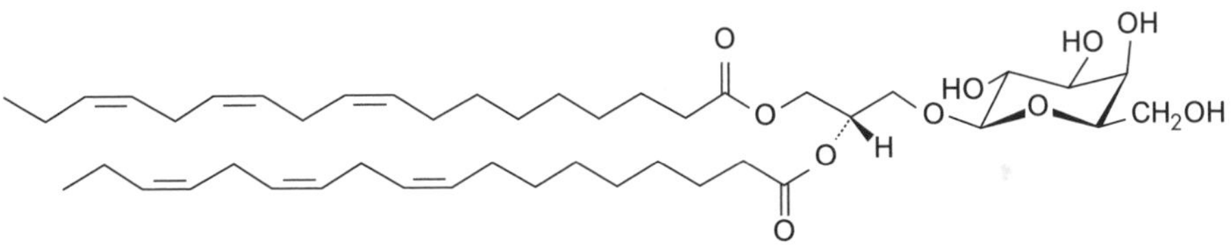

1,2-di-(9Z, 12Z, 15Z)-octadecatrienoyl-3-O-β-D-galactosyl-sn-glycerol

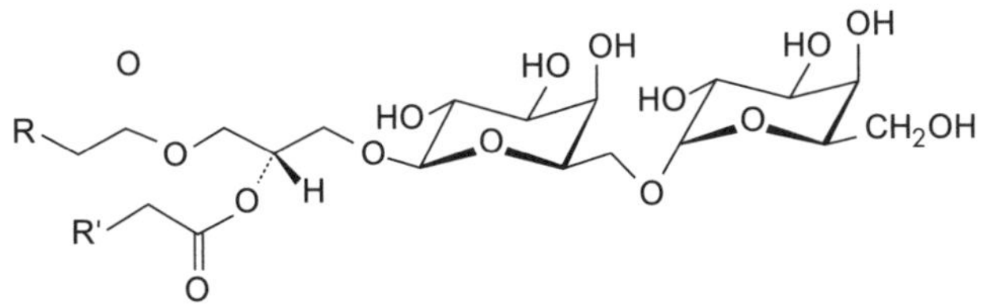

di-galactosyl-diacylglycerol

Figure 23 Main galactosyl-diacylglycerols.

Two main galactosyl-DAG are shown, differing in their polar heads. In contrast to glycero-phospholipids, galactosyl-DAG are neutral entities. Another difference is the presence of two polyunsaturated fatty acids at position *sn*-1 and *sn*-2 of this glycero-lipid. This double polyunsaturation has obvious fluidifying consequences in plant chloroplasts membranes that contain those amphiphilic components. The example of two linolenoyls in the same molecule is shown.

Galactosyl-diacylglycerols mainly differ from glycero-phospholipids by making uncharged parts of membranes, with polar heads consisting of sugars instead of phospho-choline, -ethanolamine, -serine, etc.

abundant as major components of chloroplast membranes of green plants. This make them a major lipid class on earth. Their peculiarity is to be highly unsaturated with the two *sn*-1 and *sn*-2 fatty acyls being polyunsaturated acyls corresponding to first linoleate and second linolenate for relative abundance (Fig. 23). This must provide high fluidity to membranes rich in these double PUFA lipids. The main species are mono-galactosyl-diacylglycerols (MGDG) and di-galactosyl-diacylglycerols (DGDG) (Fig. 23).

2.2 Lipid antioxidants and polyisoprenes

Natural antioxidants are quite numerous and classically separated into lipid-soluble and water-soluble molecules. This paragraph will be limited to the most well-known lipid-soluble vitamins and carotenoids of biological relevance.

Tocopherols are very hydrophobic molecules with a phenol group acting as a scavenging antioxidant. When the hydrogen radical of this group is given to arrest the propagation of peroxidation (see the cascade in Figure 19) the tocopheryl radical can be stabilized as a semi-quinone and with the opening of the adjacent oxane ring. Otherwise, the tocopheryl radical may be reduced back into tocopherol by vitamin C/ascorbic acid (two tocopheryls reduced by one vitamin C) which becomes dehydro-ascorbate. The hydrophobic part of tocopherols, in particular the saturated isoprenoid (three isoprene units) tail allows the molecule to deeply anchor itself within cell membranes. As such, tocopherols are good antioxidants to protect against membrane lipid peroxidation.

Tocopherols differ with the number of methyl groups on the phenol cycle. The two main biologically relevant ones, α- and γ-tocopherols, are represented.

As indicated in its name, tocotrienol is a triene with one double bond per isoprene unit. It is generally assumed that this triene increases the antioxidant power of the molecule compared to tocopherols. Alpha-tocotrienol represented in Figure 24 is found in substantial quantities in crude or red palm oil.

Ubiquinol, the reduced form of ubiquinone, is also well-known as a natural antioxidant. Ubiquinol is the most abundant quinol, with a very long isoprenoid (ten isoprene units), and thus called Coenzyme Q_{10}. Apart from its antioxidant activity under the quinol form, ubiquinol is mainly recognized as a key intermediate in respiratory chains within inner mitochondrial membranes, where it transfers both electrons and protons.

The most relevant natural lipid-soluble antioxidant is vitamin E which includes several tocopherol isomers. The most abundant is α-tocopherol, followed by γ-tocopherol (Fig. 24). Tocopherols are present in most cells to serve as endogenous antioxidant, protecting against peroxidation of membrane PUFA. They are also transported in blood plasma, bound to high- and low-density lipoproteins. In addition to be antioxidants, α- and γ-tocopherols exhibit slightly different properties, such as a positive action of α-tocopherol on fertility, and apoptotic activity of γ-tocopherol.

Plant oils, e.g. palm oil, contain some tocopherols in which the poly-isoprene chain is unsaturated, which increases their antioxidant potential. They are called tocotrienols.

Also, ubiquinone, especially coenzyme Q_{10}, is well distributed as it is a required component of respiratory chains located in mitochondria inner membranes. It is also present in blood plasma lipoproteins where it plays a recognized role as antioxidant in its quinol form (Fig. 24).

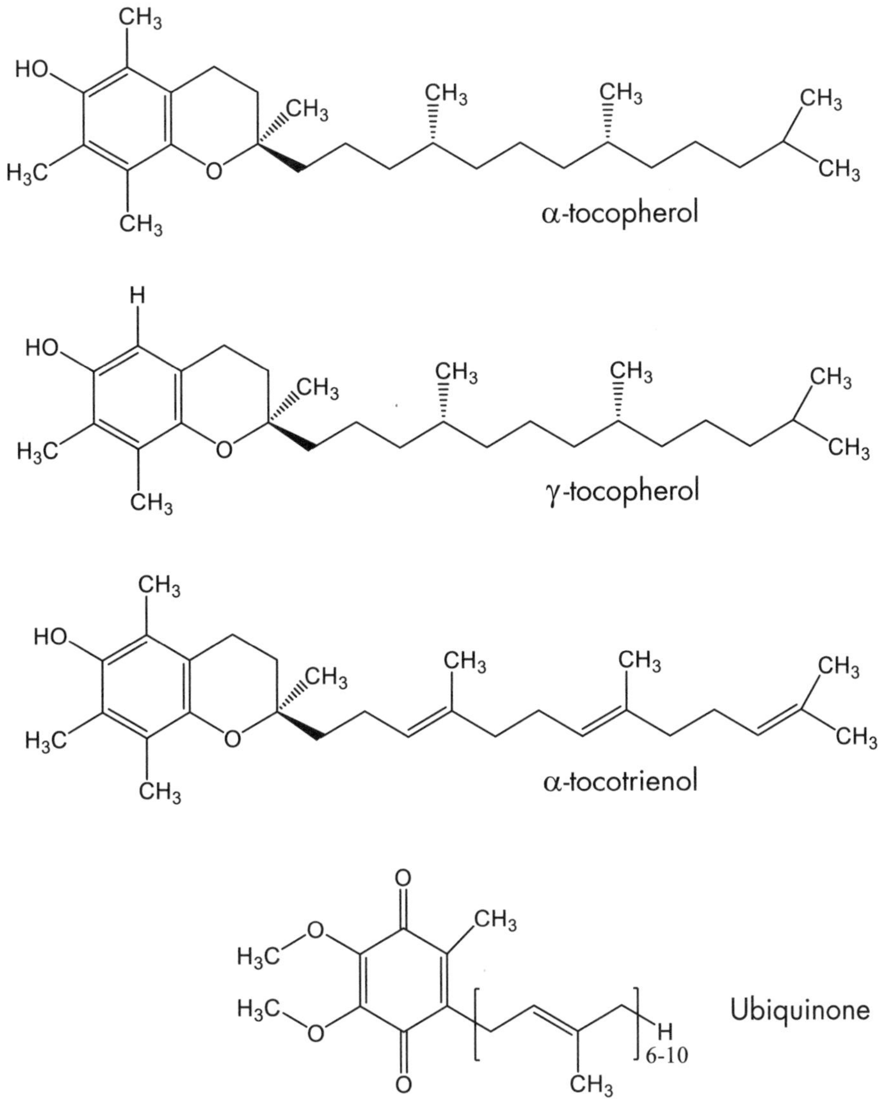

Figure 24 Main tocopherols/tocotrienols and ubiquinol.

Other lipid molecules exhibiting antioxidant activity by trapping reactive oxygen species (ROS) are carotenoids. The most abundant and relevant ones are carotenes themselves, especially β-carotene, which is the precursor of vitamin A, and lycopene (Fig. 25). They are abundant in carrots and tomatoes, respectively. Some xantophylls of biological interest are lutein and zeaxanthin (Fig. 25), also abundant in plants and eggs; they accumulate into the macula to protect retina against ROS.

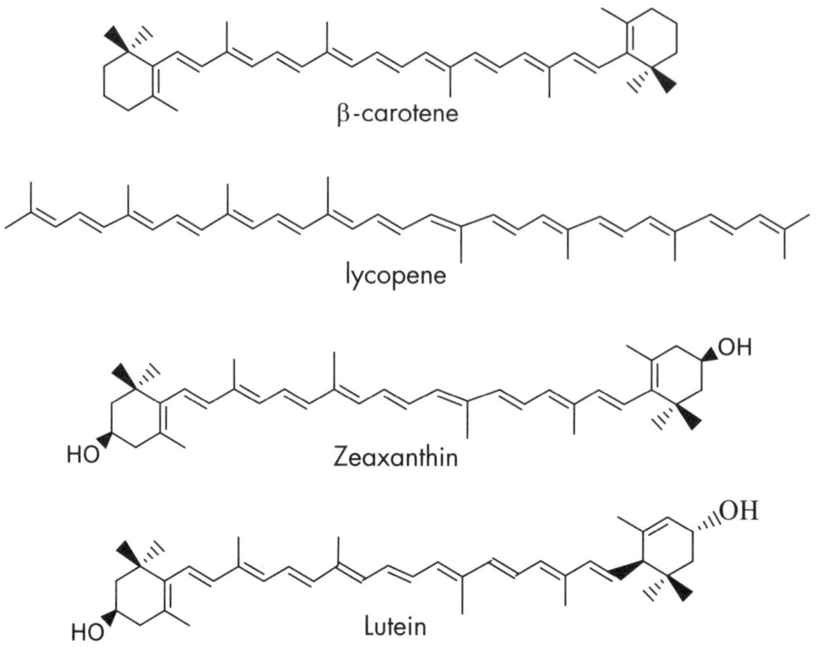

β-carotene

lycopene

Zeaxanthin

Lutein

Figure 25 Some carotenoids and xanthophylls.

Carotenoids are highly unsaturated (eleven conjugated double bonds) distributed within eight isoprene units. The most abundant carotenoid in carrots is β-carotene (Fig. 25). Lycopene, another abundant polyisoprene with eight isoprene units, is abundant in tomatoes. It looks like carotene with opened end rings. The antioxidant activity of carotenoids is linked to the high number of conjugated double bonds.

Xantophylls like lutein and zeaxanthin (major xanthophylls accumulating in the retina) are carotenes with mono-hydroxylated end cycles. Their recognized antioxidant activity is unlikely to depend on these hydroxyl groups that do not have the property of phenols, but to their high unsaturation like in carotenoids.

Other poly-isoprenes have totally different activities. Vitamins K contain a naphto-quinone core with a poly-isoprene chain (Fig. 26). Vitamins K play a crucial role as co-factors in post-translational modifications of proteins that must chelate calcium ions (e.g. some clotting proteins such as prothrombin). The modification corresponds to γ-carboxylation of glutamate residues.

The last example given in this paragraph is that of geranyl and farnesyl residues (Fig. 26), which may be combined in multiple ways to make poly-isoprenes, and are used to make other post-translational modification called prenylation, as it occurs in oncogene proteins. The most frequent protein prenylations are farnesylation and geranyl-geranylation.

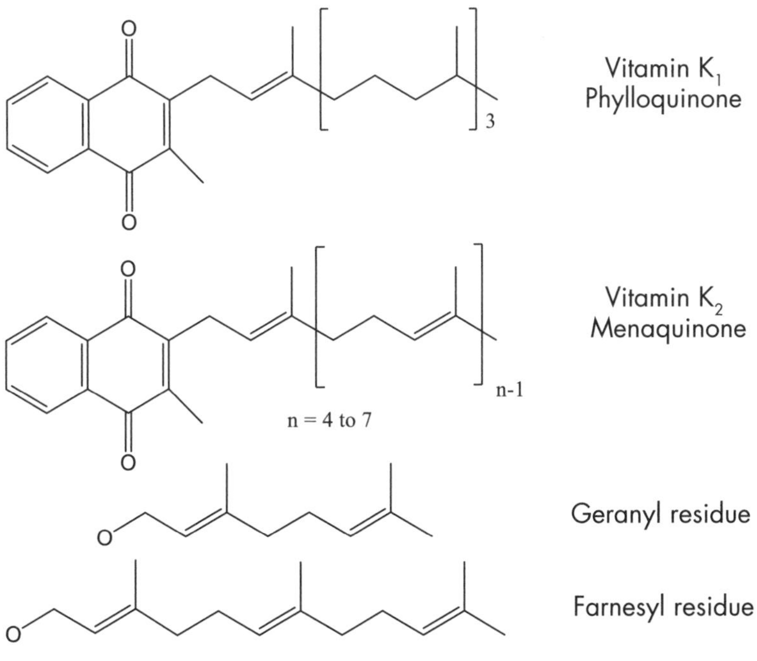

Vitamin K_1
Phylloquinone

Vitamin K_2
Menaquinone

n = 4 to 7

n-1

Geranyl residue

Farnesyl residue

Figure 26 Vitamins K and geranyl/farnesyl residues.

Vitamins K are methyl-naphtoquinone with an isoprenoid tails made of different number of isoprene units. The quinone part of the molecule plays a crucial role in the glutamate γ-carboxylation of proteins, like those involved in chelation of calcium ions. The best known example belongs to blood coagulation with prothrombin that must be 10 to 12 times γ-carboxylated within its N-terminus end to be efficiently converted into thrombin for cleaving fibrinogen into fibrin and making the clot. As a result, efficient anti-clotting therapeutic drugs (e.g. coumarin derivatives) are anti-vitamins K.

Geranyl and farnesyl residues are poly-isoprenes (two and three isoprene units, respectively) mostly used in post-translational modification (lipidation/prenylation) of proteins. They make thio-ether bonds with cysteine residues at the n-3 position of the C-terminus end of target proteins, which is followed by removal of the amino-acid residues next to the carboxyl end of the modified cysteine, and finally methylation of the new free carboxyl end into methyl-ester.

2.3 Endocannabinoids

Endocannabinoids are lipid-derived, mainly from arachidonic acid, small molecules activating cannabinoid receptors, more or less mimicking the action of the main phyto-cannabinoid, tetrahydrocannabinol. They work in a retrograde signaling process (from postsynaptic to presynaptic cells), when compared to

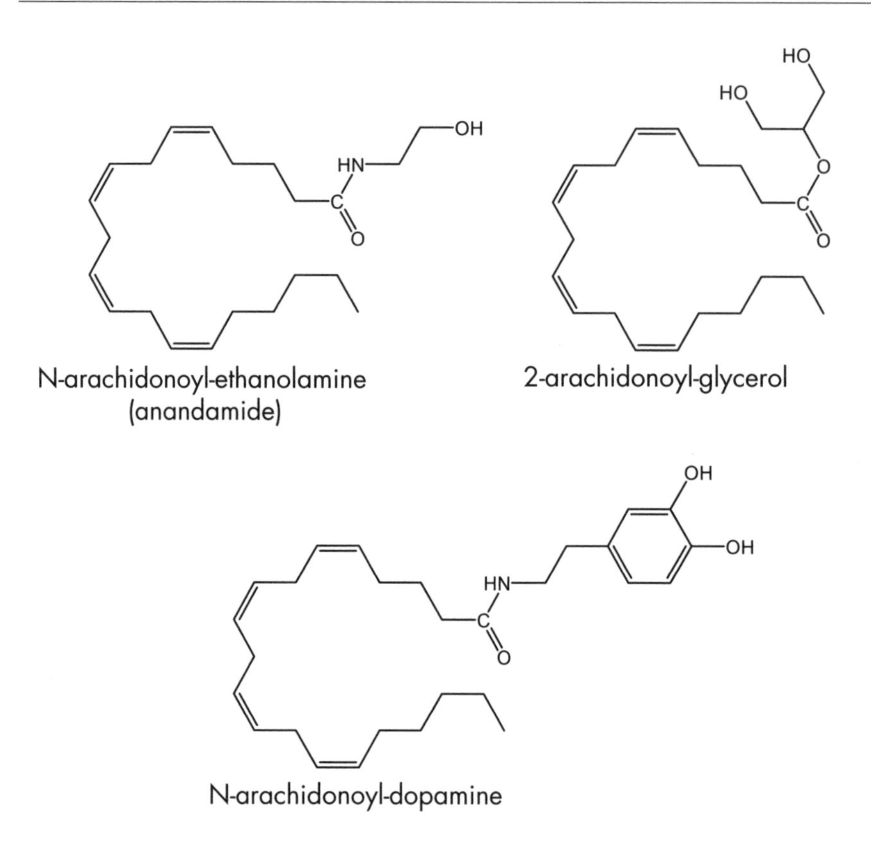

Figure 27 Three main endocannabinoids based on arachidonic acid.

The first endocannabinoid ever described has been anandamide or N-arachidonoyl-ethanolamine or arachidonoyl-ethanolamide. Its biosynthesis is very peculiar as it starts by conjugation of ArA with the ethanolamine moiety of phosphatidylethanolamine. The source of ArA seems to be 1-arachidonoyl-2-acyl-GPL (mainly PC), a molecular species that must be quite transient because this type of phospholipid with ArA at the *sn*-1 position is not common at all. Nevertheless, a specific transferase acts upon this phospholipid species to make N-arachidonoyl-phosphatidylethanolamine. Then a specific phospholipase D cleaves the conjugate to release anandamide and phosphatidic acid.

The second relevant endocannabinoid, 2-arachidonoyl-glycerol (2-AG) is made much more easily. Indeed, ArA-containing GPL hydrolyzed by PLC, especially PIs, will produce 1,acyl,2-arachidonoyl-glycerol (DAG) which can be cleaved into 2-AG by DAG lipase (see Figure 1). However, the stability of this endocannabinoid is questionable due to the easy migration of the arachidonoyl moiety to *sn*-1 or *sn*-3 position.

The third endocannabinoid, however much less abundant than the first two, is N-arachidonoyl-dopamine. Up to now, the mechanism to produce it has not been characterized yet.

neurotransmitters. The first one to be described is N-arachidonoyl-ethanolamine (anandamide). Another important endocannabinoid is 2-arachidonoyl-glycerol (2-AG) (Fig. 27). Other endocannabinoids have been described, although their biological relevance is not so obvious. Among those, N-arachidonoyl-dopamine is worth to be mentioned as dopamine is an important neuro-mediator (Fig. 27). Some analogs of the two most representatives, anandamide and 2-AG, namely virhodamine as an ester of arachidonic acid and ethanolamine, and noladin as the 2-arachidonyl-glycerol ether, are other cannabinoid agonists found in substantial amounts compared to the corresponding amides.

Interestingly, the lipid moiety of those endocannabinoids (arachidonoyl residue) can be converted into prostaglandins named prostamides. The most popular one is $PGF_{2\alpha}$-ethanolamide (called bimaprost), known for ocular hypotensive effects. Considering the lower specificity of lipoxygenases, compared to cyclooxygenases, both the endocannabinoids anandamide and 2-AG may be easily oxygenated by lipoxygenases.

Other fatty acid residues may replace the arachidonoyl moiety, such as oleoyl-ethanolamide and docosahexaenoyl-ethanolamide, the latter being recently named synaptamide because of its synaptogenic effects.

Whereas 2-acyl-glycerols are easily made from diacylglycerols by diacylglycerol cleavage (see Di-acyl-glycerol, phosphatidate and lyso-phospholipids, pp. 18-20), N-acyl-ethanolamides such as anandamide and synaptamide are produced by quite original processes which have been described in detail for anandamide. Regarding anandamide, arachidonoyl at the *sn-1* position of a glycerophospholipid is transferred to the polar head of phosphatidyl-ethanolamine to make the N-arachidonoyl link. Then, a specific phospholipase D (different from those described in Lipolytic enzymes involved in the release of bioactive lipids, pp. 11-18) separates anandamide from the remaining phosphatidic acid. This process must be tightly controlled as *sn-1* arachidonoyl is not abundant, and believed to be transient.

2.4 Sphingolipids

Sphingolipids are all based on the presence of the sphingosine or sphinganin residues, which have hydrophobic properties, and are derived with a fatty acid, usually saturated or at most mono unsaturated, to make an amide with the primary amine of the sphingoid base. In the case of sphingosine, the condensation product, N-acyl-sphingosine, is called ceramide. The primary alcohol of ceramide is derived with phospho-choline (sphingomyelin), or single sugars, or oligosaccharides (glycosphingolipids), to make the polar head (Fig. 28). This polar head made of oligosaccharides is of various carbohydrates, and plays crucial roles in cell-cell recognition. A very common example is that of ganglioside M1 (GM1) (Fig. 28) that is present in most cell membranes.

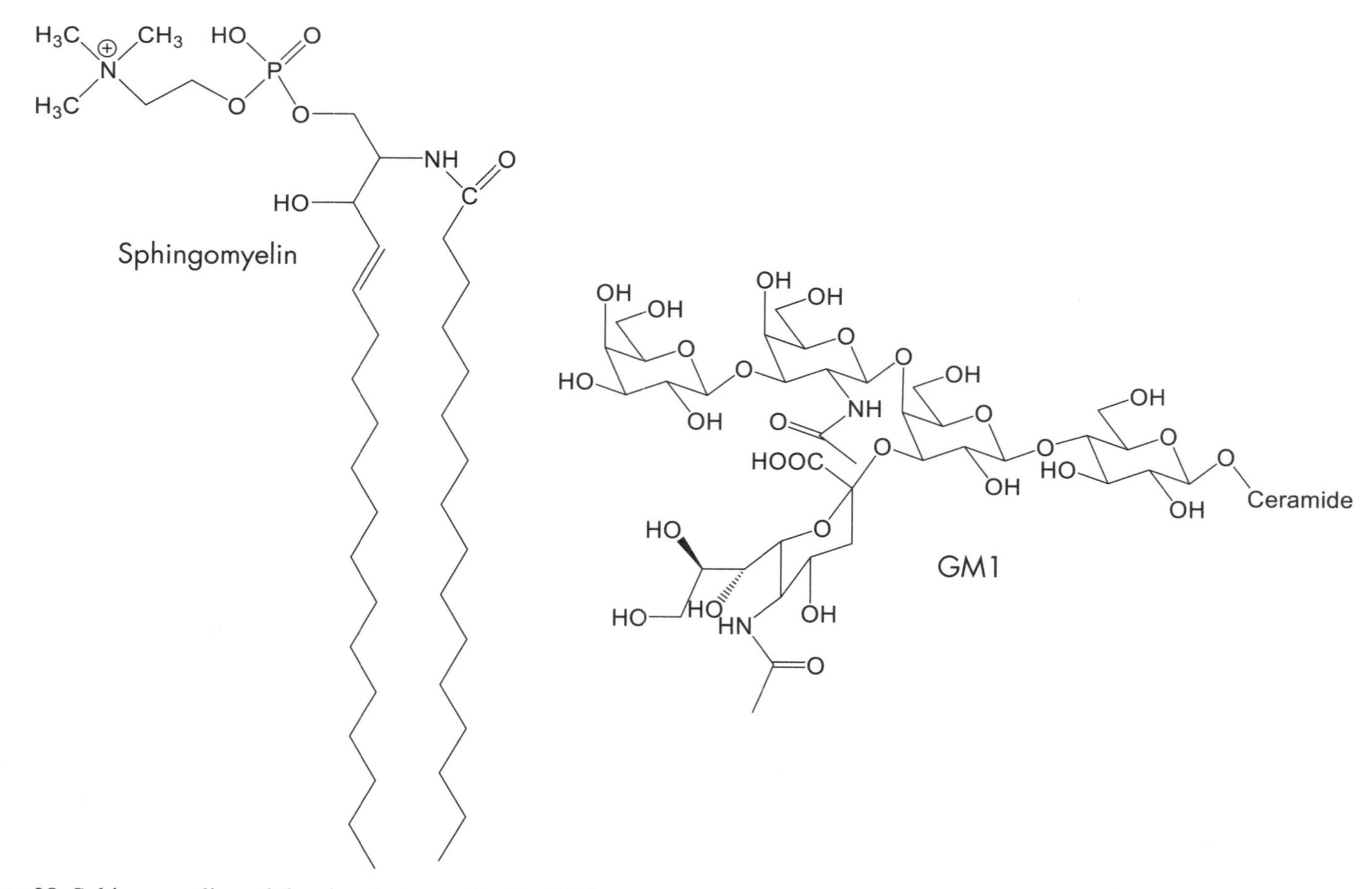

Figure 28 Sphingomyelin and the abundant ganglioside GM1.

As phosphatidylcholine, sphingomyelin is a choline phospholipid, but the hydrophobic tail is made of the acyl amidifying the primary amine, and of the sphingosine moiety. Most of the time, the N-acyl is a saturated one, which makes very packed the double chains for the hydrophobic tail of all sphingolipids, providing that the sphingosine residue has one but *trans*/E double bond.

The polar head of glycosphingolipids is made of carbohydrate residues instead of phospho-choline as in sphingomyelin. Their polar head are sometimes made of a big oligosaccharide attached to the ceramide moiety (N-acyl-sphingosine). A good example is given with GM1, a relevant so-called ganglioside because it contains at least one sialic acid residue. GM1 is a frequent marker of lipid rafts as a receptor of cholera toxin.

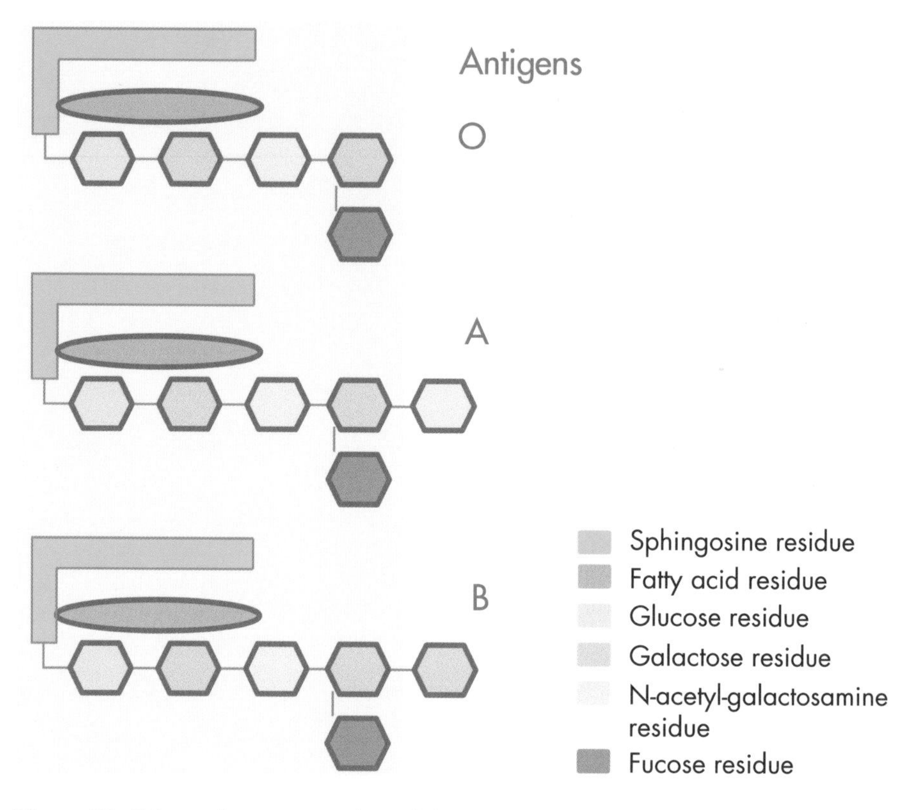

Figure 29 Schematic representation of the well-known A, B & O blood groups.

The well-known A/B/O blood groups are frequently typical glyco-sphingolipids out of the big family of gangliosides (e.g. GM1 shown in Figure 28). They only differ by the structure of their oligo-saccharide moiety. Indeed, the addition of N-acetyl-galactosamine to the penta-saccharide that makes the O antigen, makes the A antigen. In the B antigen, N-acetyl-galactosamine is replaced by galactose.

Each oligo-saccharide is attached to the membrane by the anchored ceramide.

Other well-known glycosphingolipids of biological interest are the A, B and O blood groups. They are made of penta- or hexapolysaccharides attached to the ceramide moiety with an only variation of one carbohydrate: an additional N-acetyl-galactosamine or galactose residue to the O antigen, making A or B antigen, respectively (Fig. 29).

The hydrophobic part of sphingolipids is made of N-acyl-sphingosine. The acyl moiety is most frequently saturated (sometimes mono-unsaturated), so that the only *E/trans* double bond of the sphingosine moiety, in conjunction with the saturated N-acyl, confers viscosity to sphingolipids. As such, they are the main components, with cholesterol, of lipid rafts in biological membranes, which are believed to act as signaling platforms.

2.4.1 Lipolytic enzymes acting upon sphingolipids

Enzymes which degrade sphingomyelin are called sphingomyelinases. They work like phospholipases C by cleaving the molecule between the hydrophobic moiety (ceramide residue) and phospho-choline. Sphingomyelinases are essentially classified according to their optimal pH such as acidic, alkaline, and neutral. Although not well described, a glycosphingolipid-splitting enzyme (ceramide-glycanase) has been reported as able to release the oligosaccharide moieties, whatever their nature.

2.4.2 Sphingomyelin and derived products

Sphingomyelin is the only sphingosine-containing phospholipid. In contrast to glycero-phospholipids which include classes and subclasses, sphingomyelin has only one polar head made of phospho-choline and few variations in the N-acyl group (mainly 16:0, 24:0, or 24:1). Consequently, the number of molecular species is quite limited. The structural function of sphingomyelin in membrane is known for years, which is partly in accordance with its rigidifying effect in lipid rafts (see above). However, sphingomyelin can generate a series of products by sequential enzymatic cleavages (Fig. 30). The first step, using sphingomyelinases, leads to phospho-choline and ceramides, the latter being counterparts of diacylglycerols from glycero-phospholipids. Ceramides are important bioactive molecules, by promoting the apoptosis process. In a second step, ceramides can be hydrolyzed to release the acyl moiety and sphingosine. Sphingosine is known as an inhibitor of protein kinase C, however, this is hard to tell whether sphingosine concentrations produced *in situ* are biologically relevant for such an effect. Finally, sphingosine may be phosphorylated on its primary alcohol to release sphingosine-1-phosphate (Fig. 30), a potent mitogenic molecule believed to be a relevant actor in cancero-genesis. The phenotype of the cell will then drastically differ according to the predominance of the degradation product.

2.4.3 Glycosyl-ceramides

The very diverse family of glycosyl-ceramides has in common the ceramide moiety with few variations in the N-acyl group. The vast molecular differences relate to the polar head, which consists of various sugar residues in a myriad of oligosaccharides.

Among the numerous glycosyl-ceramides are gangliosides, of which the oligosaccharide moieties contain at least one sialic acid (neuraminic acid) residue. Then subfamilies are GM, gangliosides with one sialic acid residue, GD with two, GT with three, and so on. Gangliosides play important roles in nervous tissues. GM1 (shown in Figure 28) is also the well-known receptor of cholera toxin, and it is located in lipid rafts. GM2, which accumulates in the

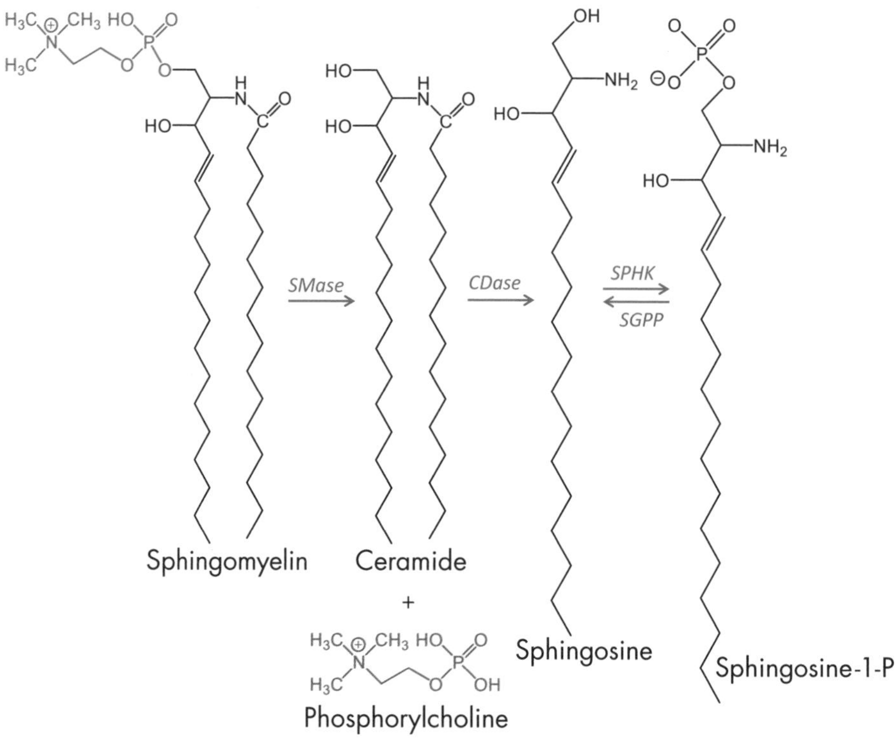

Figure 30 Sequential degradation of sphingomyelin into bioactive derivatives.

neuro-disorder called Tay-Sachs disease, derives from GM1 by removal of the end galactose residue, and GM3 from GM2 by removal of the new end N-acetyl-galactosamine. All gangliosides are important components of nerve cells, but they are also present in the periphery. This is the case of GM1 (already mentioned in lipid rafts) and GM3 which is related to insulin resistance in adipocytes as insulin-sensitive cells.

As already introduced above, O/A/B blood groups are made of five or six sugar residues, differing from only one residue (like GM1 to GM3) attached to a ceramide anchored in the red blood cell membrane (Fig. 29). The biological relevance of A/B/O antigens is quite obvious in immuno-hematology.

2.5 Steroids

Steroids derive from sterols, of which the most abundant are found in plants. However, the main animal sterol, cholesterol, and its numerous derivatives (e.g. steroid hormones, vitamin D_3, bile acids), deserve specific paragraphs because of their highly significant biological relevance.

Although sphingomyelin was considered as a stable component of cell membranes, sphingomyelinases (SMases) for its degradation are known for a long time. They are phosphodiesterases releasing phospho-choline as do phospholipases C from phosphatidylcholine. The resulting hydrophobic product is ceramide that looks like 1,2-DAG. As a matter of fact, ceramide is phosphorylated as well as 1,2-DAG by DAG kinase.

Ceramide hydrolase or ceramidase (CDase) has been characterized more recently, leading to sphingosine that can be further phosphorylated into sphingosine-1-phosphate (S-1P) by sphingosine kinase (SPHK). This end-product of the degradation sequence is water soluble enough to act within extracellular media. It is considered as a potent mitogenic agent acting through G-proteins coupled to membrane receptors, partly common to those of Lyso-PA (see Figure 5). S-1P can be hydrolyzed into sphingosine by a specific S-1P phosphatase (SGPP).

2.5.1 Sterols

All sterols and derivatives contain the cyclopentano-perhydro-phenanthrene structure (Fig. 31), except vitamins D (from cholesterol and ergosterol) in which the B cycle is opened by UV radiation in the process of their formation, and at least one hydroxyl group as a secondary alcohol. Cholesterol contains this minimal structure with two additional methyl groups between cycles A & B and C & D, plus a C_8 branched lateral chain (Fig. 31). Only one double bond is present in cycle B. Three other sterols present in plants and fungi may be mentioned. They are ergosterol, sitosterol, and stigmasterol. Compared to cholesterol, they have additional carbons on the lateral chain, and additional double bonds in both the lateral chain (ergosterol and stigmasterol) and cycle B (ergosterol) (Fig. 31).

At the opposite of plants which contain several sterols as mentioned above, animals contain only cholesterol. This molecule is susceptible to several types of chemical modifications, mainly oxidative ones (see below), but it is in great majority present in its intact form in animal tissue. As briefly mentioned in Sphingolipids (p. 59-62), cholesterol and sphingolipids are important components of plasma membranes of animal cells, making lipid rafts that are viscous micro-domains dispersed within what is called the fluid bilayer. According to its structure, cholesterol is highly hydrophobic, with the only secondary alcohol at carbon number 3 as a polar group. This means that the most part of cholesterol molecule is deeply embedded within the membrane, interacting with the water interface by the hydroxyl group at carbon 3. In addition to be a plasma membrane component, cholesterol undergoes an important compartmental metabolism. It is translocated from intracellular organelles (endosomes and lysosomes), where extracellular cholesterol is deposited by internalization of low-density lipoproteins, to plasma membranes where it is taken up by high-density lipoproteins to be recycled towards the liver (see Plasma lipoproteins, pp. 87-98).

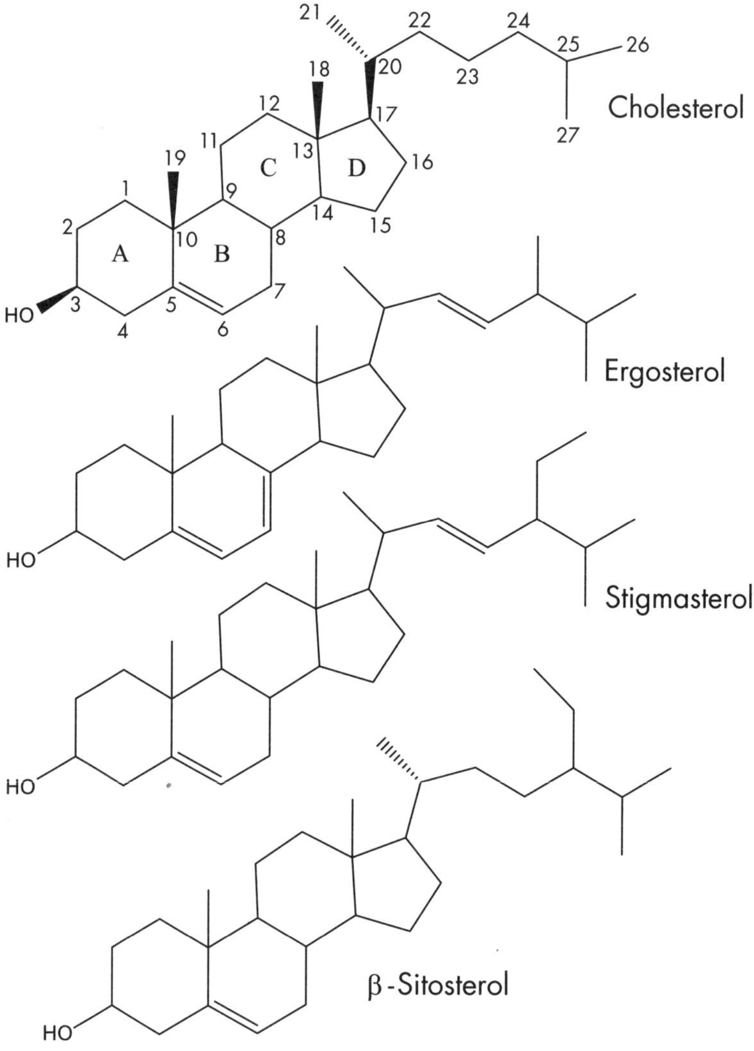

Figure 31 Structures of the main animal and plant sterols.

The main animal sterol is cholesterol while plant sterols are more diverse. They all derive from poly-isoprenes, with four cycles (A to D) and a lateral chain of different sizes and unsaturation in common. Also in common is a secondary alcohol at carbon 3 with beta stereochemistry (hydroxyl above the main plan of the molecule) and the presence of one unsaturation in cycle B.

These sterols are especially hydrophobic with only one secondary alcohol emerging from a quite hydrophobic cycles and lateral chains. Subsequently, sterols are deeply anchored within biological membranes, with the only hydroxyl group interacting with the polar environment.

2.5.2 Oxysterols

The most well-known oxysterols derive from cholesterol. Several sites of the cholesterol molecule are susceptible to oxidation either by non-enzymatic or through a cytochrome P_{450} (CYP)-dependent process. The products are usually hydroxy-cholesterols, such as 20(S)-, 22(R)-, 24(S)-, and 27-hydroxy-cholesterols synthesized through CYP actions. 25-hydroxy-cholesterol, a tertiary alcohol, is produced by a non-CYP-dependent hydroxylase (Fig. 32). The oxidation of cholesterol at carbon 7 is a prominent and more complex process. Part of 7-α/(S)-hydroxy-cholesterol is produced in the liver by a CYP-dependent hydroxylase, whereas another part, as well as 7-β/(R)-hydroxy-cholesterol, derive from a peroxidation process leading to a racemic hydroperoxide further reduced into racemic hydroxyl-derivatives. Also, two main epoxy-cholesterol, namely 5β,6β-, and 24α,25-epoxy-cholesterols are produced (Fig. 32). The former is issued from cholesterol by epoxygenation of the only double bond in the precursor in response to singlet oxygen, whereas the latter derives from squalene as an alternate pathway to lanosterol formation, so upstream to cholesterol production.

In terms of biological effects, oxysterols with the oxidized lateral chain are well-recognized as activators of liver X receptors (LXR), involved in various transcription events. Importantly, LXR mediate the transcription of hydroxy,methyl-glutaryl-CoA reductase, the key enzyme in the mevalonate formation, and target of the widely used drugs, statins. 7-hydroxy-cholesterols and their dehydrogenation product 7-oxo-cholesterol are assumed to be markers of oxidative stress and cytotoxic agents at different levels.

2.5.3 Steroid hormones

Steroid hormones all derive from cholesterol in insects and vertebrates by specific oxygenations and cleavage of the lateral chain.

For the biosynthesis of steroid hormones in mammals, the first step is the oxygenation of carbons 20 & 22 of cholesterol by cytochrome P_{450}-dependent hydroxylases, which allows the subsequent cleavage of the 20-22 bond, and the generation of pregnenolone. Pregnenolone is further dehydrogenated, and undergoes a shift of its double bond from position 7 to position 6, to produce progesterone (Fig. 33), the first bioactive hormone and key intermediate in the formation of the three well-known categories of hormone products, namely glucocorticoids, mineralocorticoids, and sex hormones. Each category requires a first oxygenation step by cytochrome P_{450}-dependent hydroxylase, sometimes common to the three pathways (such as the 17-alpha- and 11-beta-hydroxylases). The main bioactive glucocorticoid is cortisol, and the main mineralocorticoid is aldosterone (Fig. 33). The main androgen of the sex hormone cascade is testosterone, which results from the cleavage of the

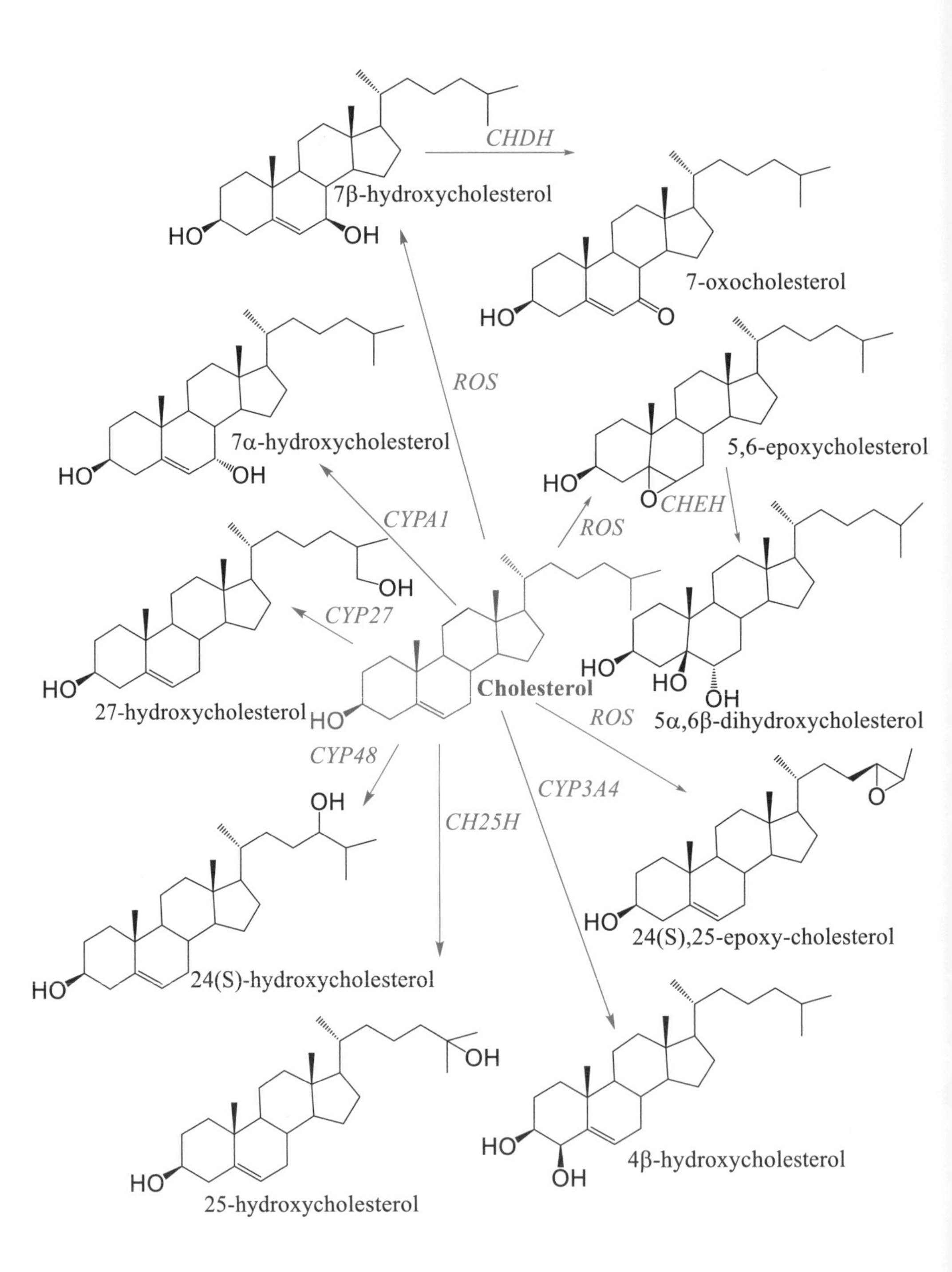

Figure 32 Chemical structures of various relevant oxysterols.

Ten different oxy-cholesterols are represented, with several of them (4β-OH-cholesterol, 7α-OH-cholesterol, 25(S)-OH-cholesterol and 27-OH-cholesterol) are known to be produced by specific cytochrome P_{450}-dependent hydroxylases. Also, 24(S)-OH-cholesterol is mainly produced from the brain cholesterol through a specific cytochrome P_{450}-dependent hydroxylase, allowing this oxysterol to cross the blood-brain barrier toward the blood stream. Another product (25-OH-cholesterol) results from the action of a non-cytochrome P_{450} hydroxylase (CH-25H). Three oxysterols result from non-enzymatic oxidation by ROS (reactive oxygen species). They are 7β-OH-cholesterol which is further dehydrogenated into its 7-oxo derivative by a specific dehydrogenase (CH-DH), and 5,6-epoxy-cholesterol made by epoxidation of the only double bond in cholesterol, further reduced into 5α,6β-dihydroxy-cholesterol by a specific epoxide hydrolase (CH-EH). The third one, 24α,25-epoxy-cholesterol, is directly produced from squalene in an alternate pathway to lanosterol.

All newly formed secondary alcohols (except 25-OH that is a tertiary alcohol) have specific stereo-chemistry in both pathways. This is expected in cytochrome P_{450}-dependent hydroxylations but more surprising in ROS-induced oxidation. The latter stereo-specificity is likely due to the steric hindrance around the cycles.

remaining lateral chain in progesterone. Testosterone is the precursor of the estrogens estradiol and estrone. It undergoes a modification of the methyl group at carbon 19 with a series of hydroxylation, double dehydrogenation of the resulting alcohol, and further decarboxylation by the enzyme complex called aromatase, meaning that cycle B of the initial cholesterol becomes aromatic after oxidation of methyl number 19 and its removal as CO_2 (Fig. 33). The product is the potent estrogen estradiol. With a redox modification (dehydrogenation of the secondary alcohol at carbon 17), estradiol is transformed into estrone that is a less potent estrogen.

2.5.4 D vitamins

D vitamins also derive from cholesterol (vitamin D_3) or ergosterol (vitamin D_2) after opening of cycle B. This means that vitamin D_3 is not a real vitamin in humans, as it can be synthesized from the endogenous component cholesterol.

Cholesterol must first be dehydrogenated in cycle B at carbon 7, which creates a conjugated diene motif (7-dehydro-cholesterol), sensitive to UVB (270-300 nm) that open cycle B between cycles A and C into pre-vitamin D_3 (Fig. 34). The product, cholecalciferol, is stabilized by isomerization, and further hydroxylated at carbons 1 and 25 in the liver and kidney, respectively, by specific cytochrome P_{450}-dependent hydroxylases into 1,25-dihydroxy-vitamin D_3 or calcitriol. As expected from their names, these cholesterol products are active on calcium ions accretion to the organism. The main active product is calcitriol, acting through nuclear receptors.

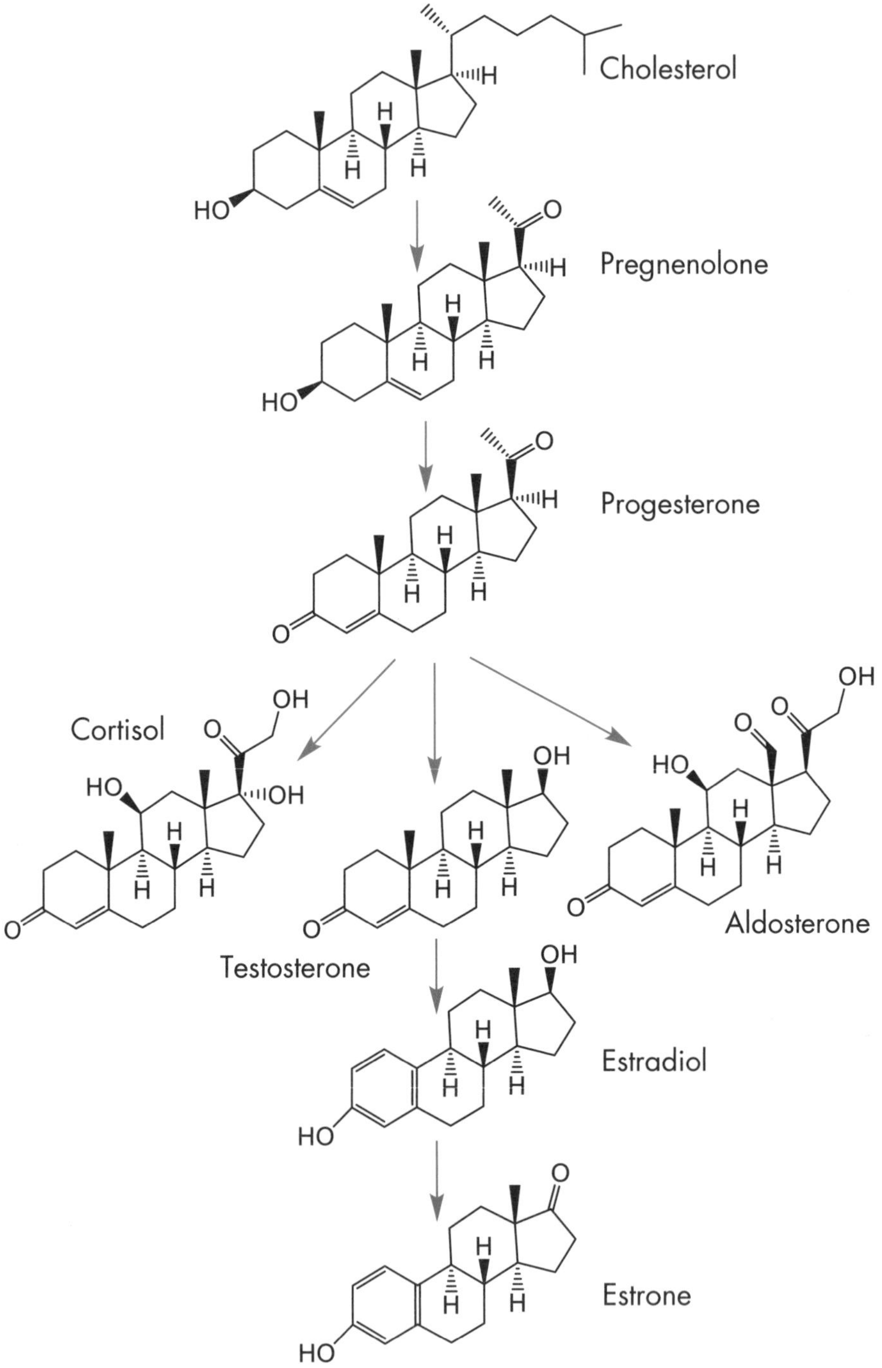

Cholesterol

Pregnenolone

Progesterone

Cortisol

Testosterone

Aldosterone

Estradiol

Estrone

Figure 33 Main pathways in the formation of mammal steroid hormones from cholesterol.

A simplified pathway for biosynthesis of the main steroid hormones in mammals is depicted in this figure.

The first steroid hormone produced from cholesterol is progesterone due to the cleavage of the aliphatic chain after hydroxylations at carbons 20 and 22. The C_{21} product, pregnenolone, is further isomerized (double bond in cycle B shifted to cycle A), which facilitates the dehydrogenation at carbon 3 into progesterone.

Progesterone is then hydroxylated at carbons 11, 17, 21 and/or 18 with the following sequences: 17α-hydroxylase, 21-hydroxylase and 11β-hydroxylase to produce cortisol, or 21-hydroxylase, 11β-hydroxylase and 18-hydroxylase with further dehydrogenation of the newly-formed 18 primary alcohol into aldehyde to produce aldosterone, the corticoid hormone that regulates the sodium-potassium exchange in the kidney with reabsorption of sodium.

Sex hormones derive first from progesterone by further cleavage of the lateral chain leading to a C19 product that is testosterone, the most potent androgen hormone. Testosterone is further metabolized by the enzyme complex aromatase consisting in first 19-hydroxylase, double dehydrogenase acting upon the newly-formed primary alcohol leading to carboxylic acid at carbon 19, and finally 19-decarboxylase accompanied by aromatization of cycle A which becomes a phenol. The final molecule is estradiol, the most potent estrogen hormone. Estrone, less potent than estradiol, is the oxidized form of estradiol at carbon 17. The oxido-reductase enzyme responsible for this oxidation may catalyze the reduction of estrone into estradiol as well, which makes this step reversible.

Vitamin D_2 is made the same way in plants from ergosterol having already the conjugated diene motif in cycle B. Thus, ergocalciferol is formed by opening of cycle B, and vitamin D_2 (Fig. 34) is further hydroxylated into the calcitriol counterpart with the same biological activity.

In addition to the well-known activity of vitamins D towards bone consolidation, recent studies have suggested beneficial effects in asthma and cancer.

2.5.5 Bile acids

Bile acids are considered as degradation products of cholesterol, made in the liver for further elimination in the feces. The main bile acids are cholic and chenodeoxycholic acids. Both of them result from β-oxidation of the lateral chain into the 24 carbon carboxylic acid, and by 7α-hydroxylase (chenodeoxycholic acid) plus 12α-hydroxylase (cholic acid) (Fig. 35). Part of them is de-hydroxylated in the intestine into deoxycholic and lithocholic acids, respectively (Fig. 35). The four bile acids then act on the digestion process by emulsifying fats. All bile acids have detergent properties, the most well-known for this action being deoxycholate that is often used as a tool for that action. Bile acids are eliminated in the feces after conjugation with glycine or taurine (e.g. glyco-cholic and tauro-cholic acids) (Fig. 35).

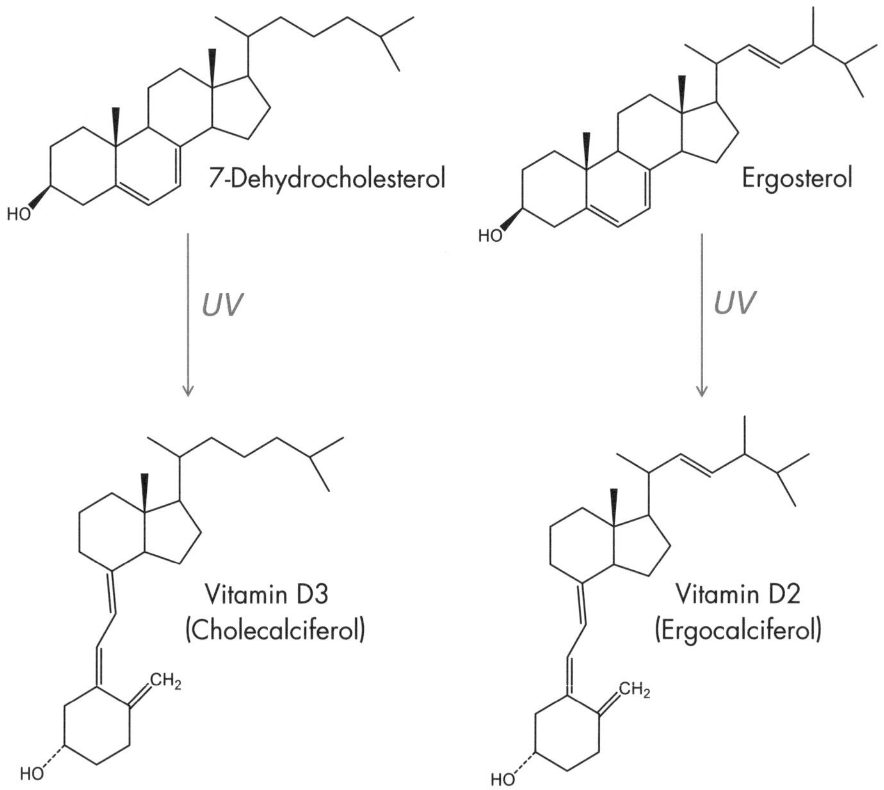

HO 7-Dehydrocholesterol HO Ergosterol

UV UV

Vitamin D3 (Cholecalciferol) CH2 HO

Vitamin D2 (Ergocalciferol) CH2 HO

Figure 34 Biosynthesis of vitamin D from 7-dehydro-cholesterol (a cholesterol derivative) and ergosterol.

Vitamins D_2 and D_3 are made the same way from ergosterol and cholesterol, respectively. The common pathway starts with the opening of cycle B when it contains two double bonds. The sigma bond between cycles A and C is fragile enough to be cleaved by UV. This means that cholesterol must first undergo creation of a second double bond (between carbons 7 & 8 in cycle B), which is done by a specific dehydrogenase.

After opening of cycle B, the products pre-vitamins D_2 and D_3 are isomerized for better stability into vitamins D_2 and D_3. D vitamins are first hydroxylated at carbon 25 in the liver to provide 25-OH-vitamins D, which are further hydroxylated at carbon 1 in the kidney. The resulting 1,25-dihydroxy-vitamins D (calcitriols) are fully active in fixing calcium ions in the body.

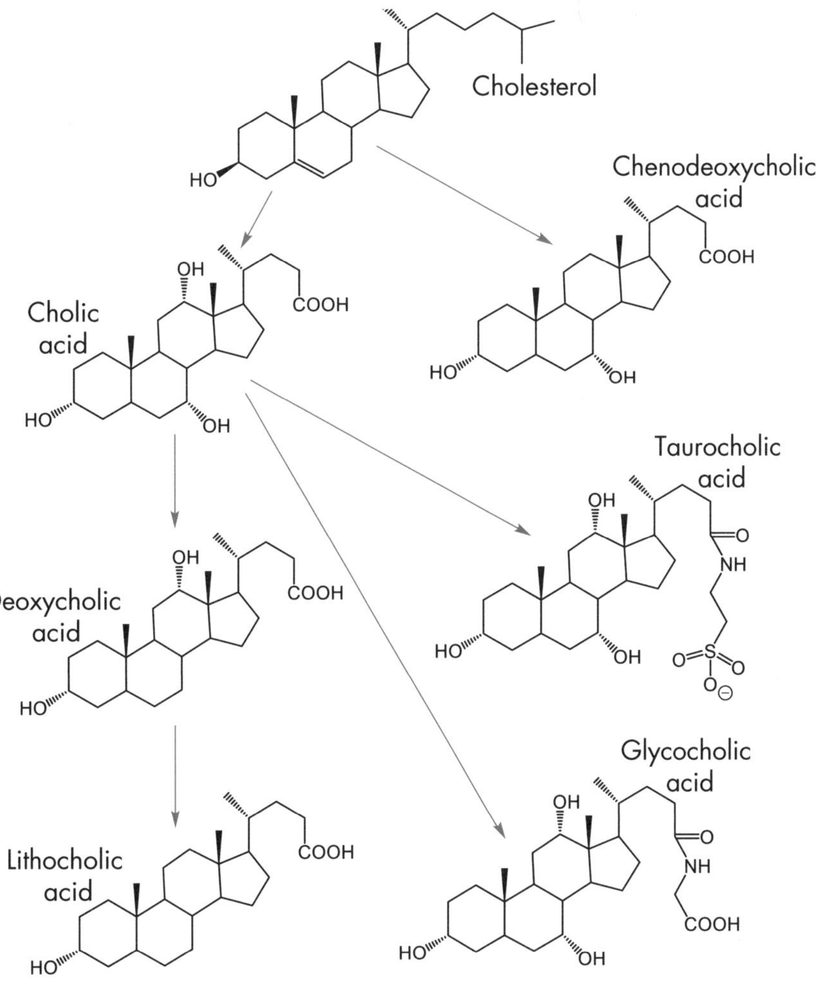

Figure 35 Different bile acids and glyco/tauro conjugates.

Bile acids, including the main ones: cholic and chenodeoxycholic acids, derive from cholesterol by shortening the lateral chain with one β-oxidation, then leading to C_{24} derivatives ending by carboxylic acid, and hydrogenation of the double bond between carbons 5 and 6. In addition, the 3β-hydroxyl in cholesterol is inverted becoming 3α, and a 7α–hydroxylase adds a second hydroxyl group. This results in chenodeoxycholic acid (3α,7α-dihydroxy-cholanoic acid). Cholic acid (3α,7α,12α-trihydroxy-cholanoic acid) results from an additional 12α-hydroxylation.

Further dehydroxylations by intestinal flora leads to deoxycholic (3α,12α-dihydroxy-cholanoic acid) and lithocholic (3α-hydroxy-cholanoic acid) acids. The elimination of bile acids is made after conjugation of the carboxylic end with glycine or taurine, making the corresponding amides glyco-cholic or tauro-cholic acids from cholic acid for instance.

3

Methodologies

As lipids are hydrophobic or amphiphilic substances, they often need to be extracted first from the aqueous biological systems by treatment with appropriate organic solvents. The lipid extracts may be then purified and/or analyzed by a variety of physical and chemical approaches with chromatographic techniques as key separation methodologies, coupled with different detection methods, including mass spectroscopy as the reference one.

3.1 Lipid extraction

To extract lipids from a biological matrix, they are first solubilized by adding an alcohol such as methanol or ethanol, which are universal lipid solvents, and precipitates the protein part of the biological sample. A solid matrix must be mechanically dispersed into the alcohol to facilitate lipid solubilization. A classical alcohol/liquid sample volume ratio is 3/1. Then, a less polar organic solvent, such as 6 volumes (referring to the initial sample) of chloroform, is added to the mixture to make a partition between the water phase and the organic phase made of chloroform/alcohol. The organic phase will contain most of the initial lipids, but the yield can be increased by a second extraction of the remaining water phase and/or by acidification according to the chemical nature of the lipids of interest. In contrast, acidification must be avoided when the extraction of plasmalogens (unsaturated sn-1 ether glycero-phospholipids) is desired, because acids will release the alkenyl moiety of those lipids.

Some technical requests must be respected for the quality of the lipid extracts. First, the extraction procedures must avoid plastic materials (culture dishes, plastic test tubes) used for biological experiments, to avoid extraction of diverse plasticizers. Second, a lipid soluble antioxidant, such as butyl-hydroxy-toluene (BHT), is highly recommended to prevent autoxidation of unsaturated lipids during sampling. Third, evaporation of organic solvents, whatever the analytical step, must be done in presence of an inert gas such as nitrogen or argon, again to prevent autoxidations.

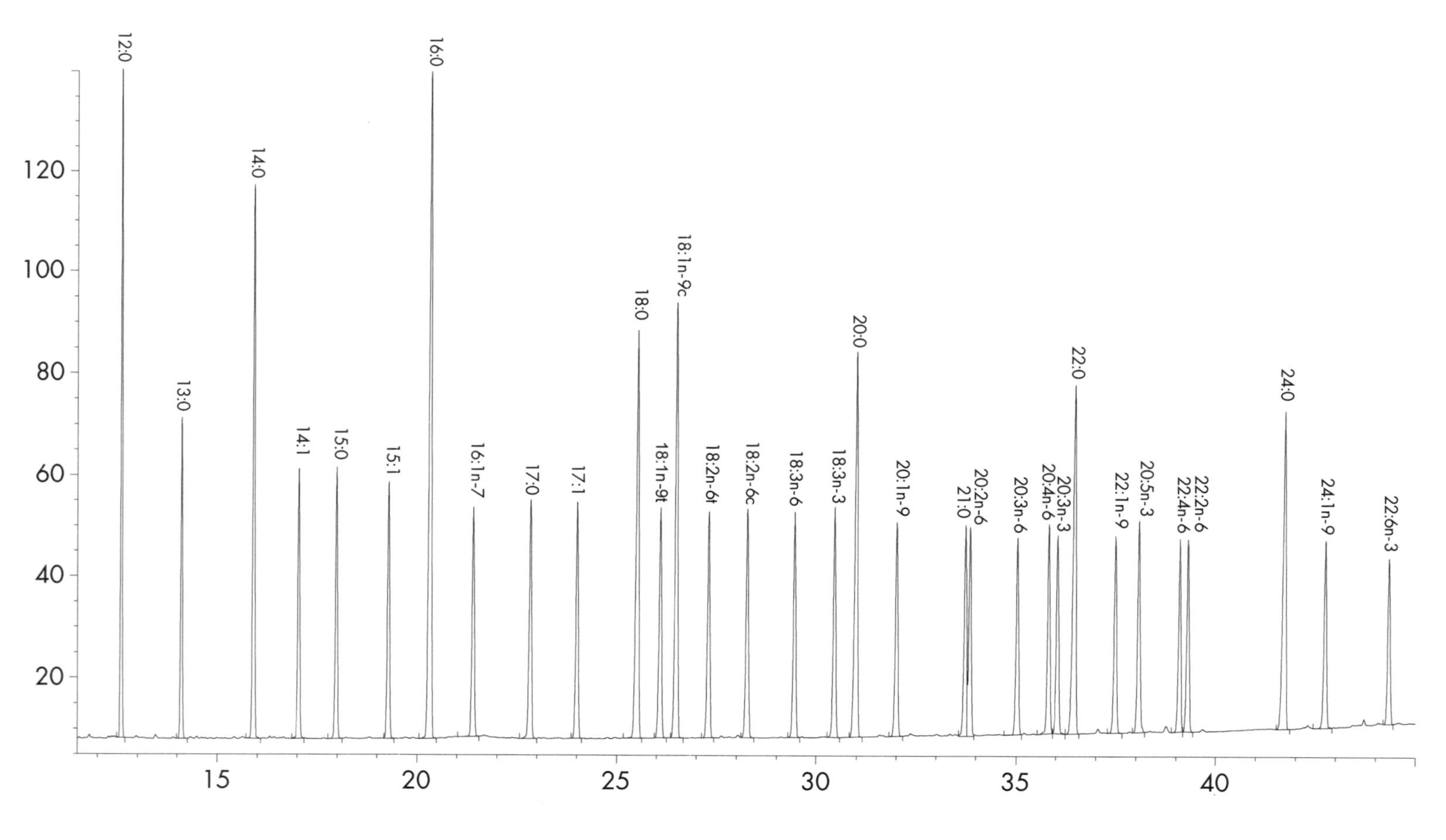

Figure 36 Chromatogram of standard fatty acid methyl esters (FAMEs) obtained by GC with a capillary column.

Standard fatty acids in methyl ester form (FAMEs) are separated by capillary gas-chromatography from a mixture of thirty-two species (from 12:0 to 22:6n-3).

FAMEs were analyzed on a 60 m i.d.: 0.25 mm BPX70 capillary column coated with 0.25 µm 70% cyanopropyl polysilphenylene-siloxane as stationary phase, using hydrogen as a carrier gas. They were detected by flame ionization. The signals shown have been obtained with the injection of 10 to 20 picomoles of each species. The low background indicates that the sensitivity can be markedly further increased.

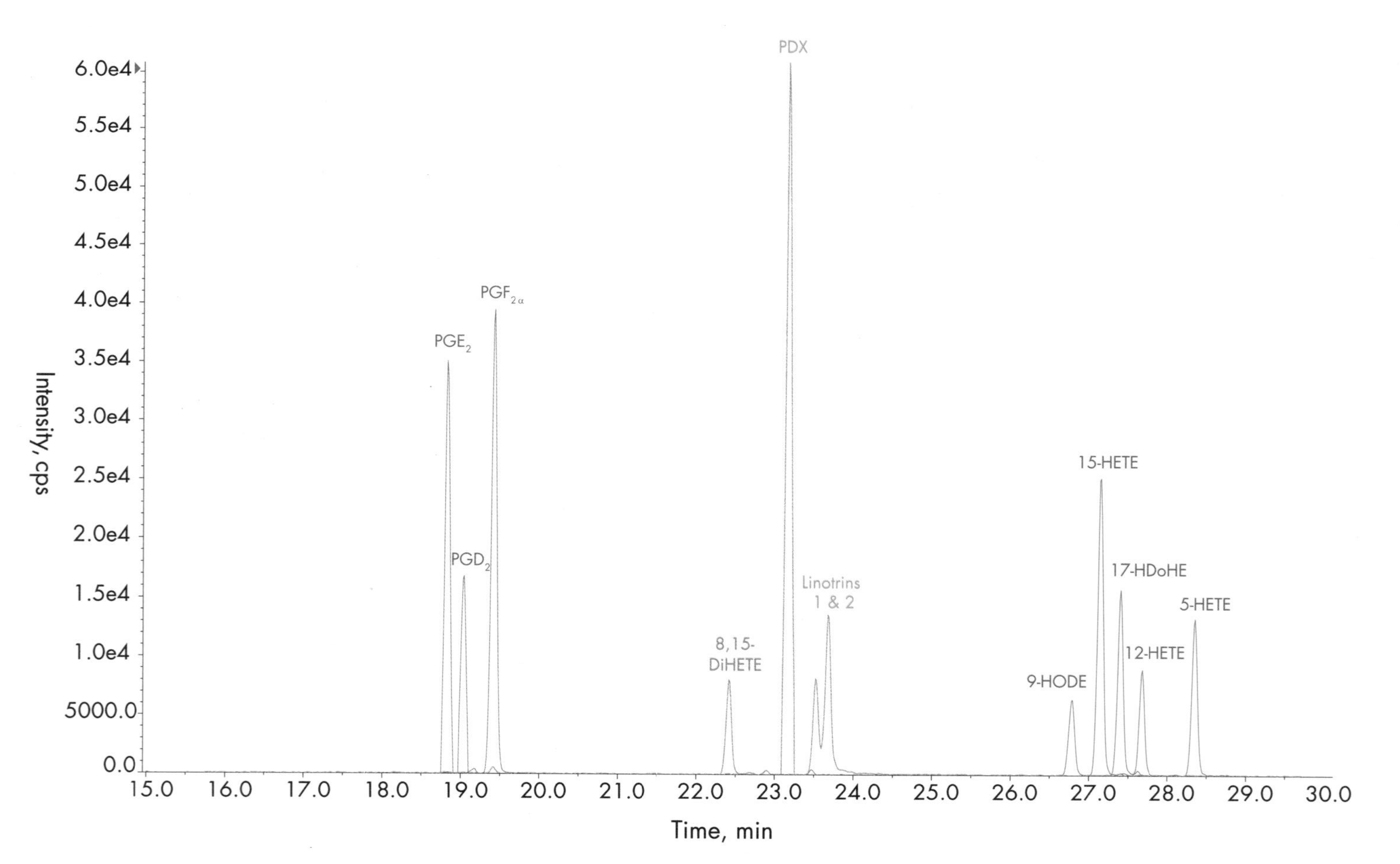

Figure 37 Liquid chromatography coupled with tandem mass spectrometry of three groups of purified oxygenated metabolites of PUFA.

They are prostaglandins (blue color), poxytrins or double oxygenated conjugated trienes (green color), and mono-hydroxylated PUFA (red color).

The Acquity UPLC® BEH C18 1.7 µm 2.1 x 150 mm column was eluted with a solvent gradient: water (pH 3): acetonitrile (90:10, v/v) (solvent A) and acetonitrile (solvent B).

The detection was based on the major ion issued from the fragmentation of each compound injected in a range of 1 to 5 picomoles.

PG: prostaglandin; PDX: 10S,17S-diOH-4*Z*,7*Z*,11*E*,13*Z*,15*E*,19*Z*-docosahexaenoic acid; 8,15-diHETE: 8S,15S-diOH-5*Z*,9*E*,11*Z*,13*E*-eicosatetraenoic acid; linotrins 1/2: 9S/R,16S-diOH-10*E*,12*Z*,14*E*-octadecatrienoic acid; 9-HODE: 9S/R-OH-octadecadienoic acid; 17-HDoHE: 17S-OH-4*Z*,7*Z*,10*Z*,13*Z*,15*E*,19*Z*-docosahexaenoic acid; 5-HETE: 5S-OH-6*E*,8*Z*,11*Z*,14*Z*-eicosatetraenoic acid; 12-HETE: 12S-OH-5*Z*,8*Z*,10*E*,14*Z*-eicosatetraenoic acid; 15-HETE: 15S-OH-5*Z*,8*Z*,11*Z*,13*E*-eicosatetraenoic acid.

3.2 Physical-chemical analyses

Lipid extracts from biological samples are usually quite complex, and need to be fractionated first prior to high resolution analysis of a given lipid family. Thin-layer chromatography is still efficient enough to do that, allowing to get total phospholipids, or phospholipid classes, triacylglycerols, steryl-esters, and some other classes of lipids.

From each class of lipids, high-performance liquid-chromatography (HPLC) or gas-chromatography (GC), coupled or not with mass spectrometry (MS), can be used for further analysis and quantitation. Regarding complex lipids, their fatty acyl composition can be classically analyzed by GC with flame ionization detection, but this first requires to transform the fatty acids into methyl-esters, which can be done in a trans-methylation process directly applied on the esters (phospholipids, triacylglycerols, or steryl-esters). A typical GC of fatty acid methyl esters is shown in Figure 36. However, when lipid species contain more than one fatty acyl (e.g. glycero-phospholipids and triacylglycerols), GC or GC-MS does not allow a molecular species analysis meaning which fatty acyl faces the other within the same molecule. HPLC-MS, and more especially with a tandem MS (HPLC-MS/MS), may provide such information.

Because many oxygenated derivatives of polyunsaturated fatty acids are doted of very specific functions, especially in cell signaling, measuring those specific oxygenated products is of primary importance. In contrast to structural lipids, those acting in lipid signaling, also called lipid mediators/messengers, are usually present in small amounts, which makes their measurement more difficult than that of structural lipids. In addition, some lipid mediators have short half-lives, so their detection only relates on their biological activities assessed by bioassays. In some cases, the half-lives are event too short to allow reliable measurement of the mediators with bioassays. The most relevant examples are those of thromboxane A_2 ($T_{1/2}$: 30 s) and prostacyclin or prostaglandin I_2 ($T_{1/2}$: 2 min). A classic and reliable alternative is to measure their immediate stable hydrolysis products, thromboxane B_2 and 6-oxo-prostaglandin $F_{1\alpha}$, respectively.

Stable lipid mediators or immediate metabolites can be measured by radio-immunoassays or enzyme immunoassays when available. Alternatively, high performance chromatography coupled with MS (e.g. GC-MS or LC-MS) may allow measuring picograms of each mediator with great specificity. LC-MS tends to replace GC-MS because the later requires prior derivatizations that make the measurements more tedious. An example of LC-MS with a tandem mass spectrometry (LC-MS/MS) is given in Figure 37.

3.3 Lipidomics approaches

As for all the "omics" approaches, there is a need for measuring as many lipids as possible, whether they are structural or bioactive, in one run or a limited number of runs. In practice, lipids can be analyzed on the basis of their chemical structures, meaning that one run corresponds to one lipid class with appropriate analytical conditions, especially their pre-separation by GC or LC.

It is worth stating that the lipidomics concept may not be restricted to the analytical matter, but may associate the enzymes responsible for the metabolism of the various lipid molecular species, as well as their genetic control. This integrated view is more functional than the pure analytical approach, and may justify focusing on one field such as structural lipidomics or mediator lipidomics.

3.4 Fluoxolipidomics

As in many fields of life sciences, extensive lipidomic analyses are not enough to properly approach the phenotypes, because of lack of kinetic data. Indeed, the half-lives of biomolecules are very different, and the resulting effects of metabolically related lipids is expected to be extremely dependent on their concentration variations, in function of time. Also, the different metabolites, occurring in function of time, often mean degradation with loss of activities. In contrast, the generation of new molecules is sometimes accompanied with different activities, especially when changing from one compartment to another. This makes those different measurements of high biological relevance, because they evaluate the fluxes attached to the various molecules.

The possible change of compartments is another parameter that affects the biological effects of those lipids. Fluxolipidomics also has the ambition of taking into consideration the effective concentrations of as many as possible lipid metabolites in a given biological compartment to facilitate the functional meaning.

The corresponding kinetics are currently not well-known, except in a few cases, but we may assume qualitative facts due to known metabolism that allows a loss or gain of biological activities. A true fluxolipidomics approach will for sure allow making a huge progress in the knowledge of lipid-dependent phenotypes. To illustrate this proposal, one may envision approaching the fluxolipidomics of essential fatty acids in humans by using stable isotope-labeled lipids. As an example, ^{13}C-labeled linoleic acid, the indispensable precursor of omega-6 fatty acids, esterified in a triacylglycerol could be ingested. Then the enrichment of various and appropriate metabolites may be measured by mass spectrometry in function of time in different blood compartments and urine. As shown in Figure 38, the isotope enrichment in linoleic acid, as well as in its main relevant fatty acid metabolite arachidonic acid, would be measured within different lipid stores (e.g. phospholipids). Also, some oxygenated linoleic acid metabolites, and especially the most representative of the numerous oxygenated arachidonic acid metabolites, would be assayed in different compartments (Fig. 38) for their ^{13}C-enrichment. According to the choice of products, the appearance of bioactive molecules and their inactive metabolites in a kinetic way should assess at best the phenotypes related to those various lipids.

Intake of ^{13}C-labeled 18:2n-6 from vegetable oil:
measurement of ^{13}C-labeled derivatives
in function of time and various compartments

Plasma: PC-18:2/PC-20:4; CE-18:2
Erythrocytes: PC-18:2/PC-20:4; PE-20:4
Platelets: PC-18:2/PC-20:4; PE-20:4
White cells: PC-18:2; PC-20:4; PE-20:4

Plasma: 9-HODE: 13-HODE;
PGD_2; PGJ_2; $15dPGJ_2$; $15dPGD_2$; PGE_2; PGA_2; PGB_2;
$PGF_{2\alpha}$; 15-oxo-$PGF_{2\alpha}$; TxB_2; 6-oxo-$PGF_{1\alpha}$; 8-epi-$PGF_{2\alpha}$;
LTB_4; 20-OH-LTB_4; 20-COOH-LTB_4; LTC_4; LTD_4; LTE_4;
5-HETE; 5-oxo-ETE; 12-HETE; 12-oxo-ETE; 15-HETE; 15-oxo-ETE

Urine: 11-dehydro-TxB_2; 8-epi-$PGF_{2\alpha}$; dinor-TxB_2; dinor-6-oxo-$PGF_{1\alpha}$

Figure 38 Choice of linoleic acid metabolites to be followed as a function of time within different compartments in a fluxolipidomics approach.

This figure enumerates several lipids for storage of linoleic acid (18:2n-6/18:2) and its main product arachidonic acid (20:4). These storage lipids are cholesterol ester (CE), phosphatidylcholine (PC) and phosphatidylethanolamine (PE).

A reasonable choice of 18:2 (HODEs) and 20:4 (PGs, LTs, HETEs,...) oxygenated metabolites is given. The list is not exhaustive but suggests products expected to show different kinetics and known to be biologically active or inactive. The comparisons between the kinetics should allow making some functional extrapolations.

Each lipid species of the list must be properly separated from the biological samples and measured for its ^{13}C enrichment by mass spectrometry. The choice shows that more than 40 species would be taken into consideration. Other metabolites issued from 18:2 and 20:4 could of course be measured, depending on the separation capacities.

In mammals, the second indispensable fatty acid is alpha-linolenic acid (ALA, 18:3n-3), precursor of the omega-3 fatty acids, gives rise to different products, especially oxygenated ones, with different activities compared to those issued from the omega-6 fatty acids (see above several paragraphs within Fatty acids and derivatives, p. 20-44). The two series being in mutual competition both metabolically and functionally, it is worth to evaluate them simultaneously. The performance of analytical tools briefly presented above (Physical-chemical analyses, pp. 75-80) allow simultaneous measurements. Figure 39 represents a choice of lipids measurable from alpha-linolenic acid after simultaneous intake of ^{13}C-linoleic and ^{13}C-alpha-linolenic acids.

Intake of ^{13}C-labeled 18:3n-3 from vegetable oil: measurement of ^{13}C-labeled derivatives in function of time and various compartments

Plasma: PC-18:3/PC-20:5/PC-22:6; CE-18:3/CE-20:5
Erythrocytes: PC-18:3/PC-20:5/PC-22:6; PE-20:5/PE-22:6
Platelets: PC-18:3/PC-20:5/PC-22:6; PE-20:5/PE-22:6
White cells: PC-18:3/PC-20:5/PC-22:6; PE-20:5:PC-22:6

Plasma: 9-HOTE; 13-HOTE; 9,16-diHOTE
PGD_3; PGJ_3; $15dPGJ_3$; $15dPGD_3$; PGE_3; PGA_3; PGB_3;
$PGF_{3\alpha}$; 15-oxo-$PGF_{3\alpha}$; delta17-6-oxo-$PGF_{1\alpha}$; 8-epi-$PGF_{3\alpha}$;
LTB_5; 20-OH-LTB_5; 20-COOH-LTB_5; LTC_5; LTD_5; LTE_5;
5-HEPE; 5-oxo-EPE; 12-HEPE; 12-oxo-EPE; 15-HEPE; 15-oxo-EPE
Linotrins; Protectins D1 & DX; Resolvins Es & Ds

Urine: 8-epi-$PGF_{3\alpha}$; dinor-TxB_3; dinor-6-oxo-delta17-$PGF_{1\alpha}$

Figure 39 Choice of alpha-linolenic acid (18:3n-3) metabolites to be followed as a function of time within different compartments in a fluxolipidomics approach.

The legend of this figure is similar to that of Figure 38, except that two PUFA metabolites of 18:3n-3, namely 20:5n-3 and 22:6n-3 are followed. As to the cyclooxygenase and lipoxygenase products from the n-6 family, the counter products of the n-3 family are shown. In addition, specific oxygenated metabolites of the n-3 family are taken into consideration, namely, linotrins (from 18:3n-3), protectins D (from 22:6n-3) and resolvins (from both 20:5n-3 & 22:6n-3). Although not exhaustive, the number of proposed molecules to be measured is higher than 60.

The separation being made between products from n-3 and n-6 derivatives, both precursors (18:2n-6 and 18:3n-3) may be ingested together.

3.5 Lipid imaging

Dynamic imaging of lipids strongly depends on the nature of these lipids, whether they are structural or mediator molecules, and depends on the availability of tracers or probes. As lipids are substantially smaller than proteins for instance, the size of probes is a crucial issue. A prudent approach is the use of labeled lipids with atoms detectable by nuclear magnetic resonance (NMR) such as carbon 13 (^{13}C), or positron emission tomography (PET) such as carbon 11 (^{11}C),. In some cases, fluoride 19 (^{19}F) or fluoride 18 (^{18}F) can be useful alternatives, respectively, when a hydroxyl group can be replaced by a fluoride atom.

More classically, a fluorescent lipid is used to localize itself, and/or to assess its movements. Bodipy-labeled lipids are available for this (Fig. 40). However, the size of the fluorescent probe is a critical issue that must be taken into consideration. Another approach to minimize this concern is the use of a fatty acyl radical that fluoresces. *Sn*-2 parinaric acyl-containing glycerophospholipids may be such an alternative (Fig. 40). In this case an additional advantage is to follow the release of parinaric acid after triggering cells with an agonist that stimulates phospholipase A_2.

The NMR approach based on natural isotopes such as ^{31}P, naturally present in phospholipids, or the use of pre-enriched ^{13}C-labeled molecules is valid if the molecule is mobile enough, which could be a limiting factor in the cell membrane bilayers. Also, the presence of natural ^{13}C and especially ^{31}P might generate a strong background, more or less covering the signal from the molecule of interest. The use of rare responder atoms like ^{19}F in substantial quantities might solve the background issue. However, adequate labeling is required to prevent any change of molecular reactivity due to the introduction of fluoride. Substitution of a "non-functionally exposed" hydroxyl group in the lipid of interest by fluoride is a promising approach.

The PET approach is based on the two coincident and opposed photons emitted at 511 Kev each after annihilation of the positron (beta +) emitted by ^{11}C- or ^{18}F-labeled lipids. In accordance to the very short period of the radioactive elements (20 and 110 min, respectively), the label must be located for an easy access part of the molecule. 1-^{11}C-labeled fatty acids have been used that way, and ^{18}F-labeled lipids can be designed as currently done with ^{18}F-2-deoxy-glucose to trace glucose in tissues.

More recent approaches use a Matrix-Assisted Laser Desorption Ionization (MALDI)-type mass spectrometry. This allows localizing lipids directly in slices of biological tissues, and providing two-dimension images. Recent promising applications have been shown for brain lipids. Alternatively to MALDI analysis, Time-Of-Flight (TOF) mass spectrometry has been successfully used.

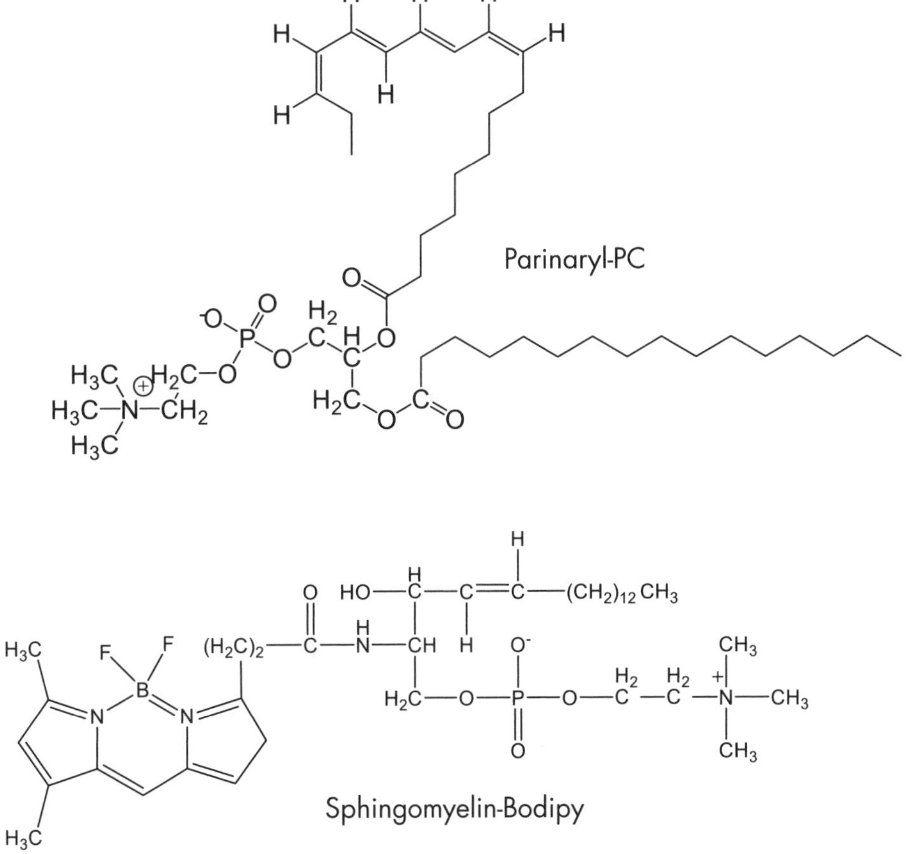

Figure 40 Structure of Bodipy-labeled and parinaric acyl-containing phospholipid.

The fluorescent dye called bodipy is dipyrro-methene substituted by a BF_2 unit. In Figure 40, bodipy is prolonged with one propionic acid allowing its conjugation with the primary amine group of the ceramide-phosphocholine moiety. This type of fluorescent probe allows investigation of cell membranes. In the case of bodipy labeled sphingomyelin, lipid rafts are especially targeted. This probe is easy to follow, but its disadvantage is its size which is quite different from the N-acyl moiety that it replaces in the phospholipid.

Another fluorescent species that mimics much more closely the endogenous acyls is parinaroyl from parinaric acid, a conjugated tetraene with the *Z,E,E,Z* geometry. The parinaroyl residue mimics enough PUFA at the *sn*-2 position of PL to be recognized by PLA_2 and released upon activation. The released parinaric acid can just be measured by fluorimetry.

4

Plasma lipoproteins

Because of their hydrophobic properties lipids cannot circulate in blood without specific hydrophilic transporters. This is well-known for classic lipid classes such as triacylglycerols, cholesterol and cholesteryl esters, and phospholipids, which are associated with specific proteins, called apo-lipoproteins, within lipoproteins. This is also valid for steroid hormones that are transported by specific binding proteins.

4.1 Lipid components

Lipoproteins are macromolecular complexes of lipids and proteins that transport hydrophobic lipids within the circulation to and from tissues. Structurally, lipoproteins are spherical complex particles made of a hydrophobic core surrounded by a hydrophilic surface monolayer. The core contains triacylglycerols (TAG) and cholesteryl-esters (CE) whereas the surface contains glycero-phospholipids, non-esterified cholesterol, and embedded apo-lipoproteins. Lipoproteins differ according to their size, density, due to their lipid and protein composition. They are usually classified into five classes with different densities of the particles according to the ultracentrifugation characteristic, namely chylomicrons, very low-density lipoproteins (VLDL), intermediate-density lipoproteins (IDL), low-density lipoproteins (LDL), and high-density lipoproteins (HDL). The total lipid content is inversely related with the density of the particle. TAG are the major lipids within the core of less dense lipoproteins (chylomicrons and VLDL), whereas they are the minor lipids of more dense lipoproteins (LDL and HDL). The latter contain instead much cholesteryl-esters in the hydrophobic core. The PL content is correlated with the density and surface area of the lipoprotein particles. Non-esterified cholesterol is present in low proportions within all lipoprotein classes, being esterified by lecithin-cholesterol acyl-transferase (LCAT) within HDL, and distributed to VLDL and LDL by cholesteryl-ester transfer protein (CETP). LCAT specifically transfers the *sn-2* acyl (mainly linoleoyl) from HDL phosphatidylcholine to non-esterified cholesterol, making CE (mainly cholesteryl-18:2), which is translocated from the surface to the core of the particles.

This is why the native discoid HDL become spherical by maturation, i.e. enriched with CE. Concomitantly, lysophosphatidylcholine (due to the loss of *sn-2* acyl groups from PC) are produced, and mainly transferred to albumin. This important metabolic step is directly associated with the reverse transport of cholesterol, a process which is crucial in preventing atherosclerosis.

Compared to those major lipid classes, non-esterified fatty acids and glycolipids are present in very low amounts in lipoproteins.

Chylomicrons, the largest and least dense lipoproteins, are mainly composed of TAG derived from dietary lipids, with a high proportion of saturated and monounsaturated fatty acyls. Chylomicrons contain only 1-2% of total proteins, being mainly apo-B48, an apo-lipoprotein exclusively found in native chylomicrons (Table 2).

VLDL also contain a high proportion of TAG, accounting for half of total lipids, and relevant amounts of CE, PL, and cholesterol. Apo-B100 is the main apo-lipoprotein of VLDL.

IDL are VLDL remnants formed after hydrolysis of VLDL TAG by lipoprotein lipase. Their lipid core contains CE and TAG in equivalent proportions, while the surface monolayer is mainly composed of PL with a low proportion of non-esterified cholesterol (Table 2).

LDL transport blood cholesterol, mainly to supply this sterol to cells. The lipid core of LDL contains around 1,600 molecules of CE and 170 molecules of TAG. The surface monolayer contains around 700 molecules of PL and 600 molecules of cholesterol. So, three quarters of LDL lipids relate to both forms of cholesterol. These values represent the average LDL, without taking into consideration size heterogeneity. Recent lipidomics analyses by LC/MS identified more than 350 different lipid molecular species from 19 lipid subclasses. Among the phospholipid classes, phosphatidylcholine (PC) is the main one, accounting for 64% of total PL. Sphingomyelin (SM) is mainly carried in the circulation by LDL, and is the second phospholipid class which mainly contains saturated fatty acyls. Lyso-phosphatidylcholine (LPC) is found in native LDL, as generated by PLA_2 or its transfer from HDL following the LCAT action. LPC is also found in oxidized LDL, following the hydrolysis of oxidized PL by platelet activating factor acetyl-hydrolase (PAF-AH). Phosphatidylethanolamine (PE) is present in minor amount, and is composed of diacyl-PE and plasmalogen PE, the latter being characterized by a vinyl ether link at the *sn-1* position and an ester link at the *sn-2* position, as stated in paragraph 2.1.2.1. Plasmalogens may contribute to the oxidation resistance of LDL, acting as an antioxidant. Phosphatidylinositol (PI) and phosphatidylserine (PS) represent minor phospholipids in LDL. Half fatty acid residues (fatty acyls) in LDL are polyunsaturated fatty acids (PUFA), namely linoleic acyl and to a lesser extent arachidonic and docosahexaenoic acyls. CE represent nearly 50% of total LDL lipids and are especially rich in linoleic acyl, accounting for 50%

total fatty acyls. All PUFA are major targets of free radical attack and may give rise to many fragmentation by-products. The LDL particles also contain lipid-soluble antioxidants to protect themselves against oxidation. The main one is α-tocopherol with 6 molecules per LDL particle. They also contain γ-tocopherol, β-carotene, α-carotene, lycopene, xanthophylls such as lutein and zeaxanthin, and ubiquinol or coenzyme Q_{10}.

Apo-B100 is the major non exchangeable apo-lipoprotein of LDL, but there is also some apo-A, E, CI, CII, and CIII. LDL may be further subdivided into large and light LDL, and small and dense LDL (Table 2). Small and dense LDL show increased TAG and decreased CE, and they are more susceptible to oxidative modifications.

HDL are the smallest and most dense lipoproteins. They constitute a highly heterogeneous group of particles, both structurally and metabolically. To date, more than 200 individual molecular lipid species have been identified within HDL. Like other lipoprotein classes, HDL are composed of PL, CE, cholesterol, and TAG, in decreasing order. PL constitute then the major lipid class of HDL, with PC being the major PL class (70%) and the richest in PUFA. SM accounts for 5-10% of total lipids, is the major sphingolipid class, and the second PL component of HDL. Lyso-PC are also found in HDL as a resulting product of LCAT activity, an enzyme primarily bound to HDL. However most lyso-PC are exported to bind other blood protein, mainly albumin, after their production by LCAT. HDL also contain lyso-sphingolipids and are the main carrier of sphingosine-1-phosphate in plasma. Considering the role of HDL in cholesterol transport through the circulation, it is worth to point out that HDL are rich in cholesterol (3-5%) and CE (15-20%) formed via the trans-esterification reaction by LCAT. CE are then very rich in linoleoyl residues, the main fatty acyl species of the PC substrate.

Recent proteomic analyses have characterized hundreds of proteins which can be divided into four groups: apo-lipoproteins, enzymes, lipid transfer proteins, and minor proteins involved in several other functions such as lipid metabolism, acute-phase responses, proteinase inhibition and complement regulation. The apo-lipoproteins present in HDL are exchangeable. The major one is apo-AI which represents 70% of total exchangeable proteins. Apo-AII is the second major apo-lipoprotein and represents 15-20% of total HDL proteins. The main enzymes are LCAT, the antioxidant paraoxonase 1 (PON1), Platelet-Activating Factor acetyl hydrolase, and glutathione peroxidase-3 (GPx-3). Lipid transfer proteins are involved in the exchange of lipids between lipoprotein particles. They are the phospholipid transfer protein (PLTP) and CETP (Table 2). HDL also contain glycosylated proteins that transport micro RNA (miRNA) and deliver them to cells, controlling gene expression.

HDL can be further divided into two main sub-fractions: large and light, lipid-rich HDL2, and small and dense, lipid-poor (protein-rich) HDL3.

Table 2 Characteristics and lipid composition of the main lipoprotein classes.

	Chylomicrons	**VLDL**	**IDL**	**LDL**	**HDL**
Density (g/ml)	< 0.95	0.95-1.006	1.006-1.019	1.019-1.063	1.063-1.21
Diameter (nm)	75-120	30-80	25-30	18-25	5-12
Electrophoretic mobility	origin	pre-β	β	β	α
Half-life	15-20 min.	4-6 hours	2 min.	2-3 days	3-5 days
Composition					
Proteins (%)	1-2	6-10	18-20	18-22	45-55
Lipids (%)	98-99	90-94	72-80	78-82	45-55
Lipid composition (% total lipids)					
Cholesteryl esters	2-4	16-22	20-30	45-50	15-20
Triacylglycerols	80-95	45-65	25-35	4-8	2-7
Phospholipids	3-6	15-20	20-24	18-24	26-32
Cholesterol	1-3	4-8	4-8	6-8	3-5
Main apo-lipoproteins	B48 (20%), A, CI, CII, CIII, E	B100, CI, CII, CIII, E	B100, CI, CII, CIII, E	B100 (90%), CI, CII, CIII, E	AI (70%), AII (20%), AIV, CI, CII, CIII, D, E

4.2 Functional dynamics

Lipoproteins allow the transport of lipids within the body. Their metabolism is open to the environment and adapted to body needs, and is divided into three main pathways: the exogenous pathway (relating to dietary lipids), the endogenous pathway (as shown in Fig. 41), as well as the pathway of reverse cholesterol transport (RCT) (shown in Fig. 42). Lipoproteins are dynamic particles that evolve within the blood circulation.

Exogenous pathway

Pancreatic lipase acts to hydrolyze triacyl-glycerols at positions 1 (last point of hydrolysis) and 3 (first point of hydrolysis). The products of this reaction are Na+ and K+ salts of fatty acids (also known as soaps). Lipid digestion by pancreatic lipases generates mono-acyl-glycerols (MAG) that are absorbed in the small intestine. Bile acids aid in this process too, forming micelles. Once inside the intestinal cells, fatty acids complex with a protein in the cytoplasm called intestinal fatty acid-binding protein (I-FABP) that increases their effective solubility and protects the cell from their detergent effects. The mono-acyl-glycerols in the digestive cells are converted back to triglycerides/triacyl-glycerols (TAG), and packaged into lipoproteins in the bloodstream called chylomicrons.

This pathway enables the provision of dietary lipids (exogenous) to target tissues for energy production, storage, or molecule synthesis. Dietary fats (mainly TAG) are hydrolyzed by TAG pancreatic lipase at the brush-border of small intestine cells into non-esterified fatty acids and MAG, as explained above, allowing their entry into the cell, before being reassembled within the cell into new TAG into the endoplasmic reticulum. Beginning in the endoplasmic reticulum and continuing in the Golgi, TAG are packaged with cholesterol, apo-lipoproteins, and other lipids into chylomicrons. These nascent chylomicrons are secreted from the intestinal epithelial cells into the circulation through the lymph, and bypass the liver circulation to target specific tissues. The fatty acid composition of chylomicron lipids (mainly TAG) is very close to the one of ingested fat. As lipoproteins are dynamic particles, they can be modified in the circulation. Indeed, HDL donate apo-CII and apo-E to the nascent chylomicrons that are then considered mature. In muscle and adipose tissue, chylomicron TAG are hydrolyzed into non-esterified fatty acids (NEFA) and mono-acyl-glycerol (MAG) by lipoprotein lipase (LPL), an enzyme attached to endothelial cells by non-covalent binding to glycosaminoglycan associated with the lumen, activated by apo-CII. MAG are hydrolyzed by MAG lipase into glycerol and NEFA that can be absorbed by peripheral tissues, especially the adipose tissue and muscles, for storage or energy production.

Within the circulation, the hydrolyzed TAG-chylomicrons are considered as chylomicron remnants, which circulate until they interact, via apo-E, with specific receptors on the hepatocyte membranes. This interaction causes the endocytosis of chylomicron remnants, which are subsequently hydrolyzed (both lipids and proteins) within lysosomes. Lysosomal hydrolytic activities releases glycerol and fatty acids into the cell, and those fatty acids can be stored or used for energy. Usually, most of the chylomicrons disappear from the circulation within 12 hours after a meal.

Endogenous pathway

This pathway transports endogenous lipids from the liver to peripheral tissues, few hours after a meal, as soon as the amount of blood chylomicrons is low, the liver provides TAG, needed to peripheral tissues. Lipids are assembled with apo-B100 inside the liver within nascent VLDL particles that are released into the circulation. As for the chylomicron metabolism, HDL particles exchange apo-CII and apo-E with mature VLDL particles. In the circulation, LPL, present at the luminal surface of endothelial cells, hydrolyses VLDL-TAG to release fatty acids and glycerol, taken up by muscle cells for energy, or by adipose cells for storage. Once processed by LPL, VLDL become VLDL remnants, most of them being taken up by the liver via remnant receptors. The remaining remnant particles become IDL, smaller and denser lipoproteins. Some of the IDL particles are reabsorbed by the liver via receptor-mediated endocytosis,

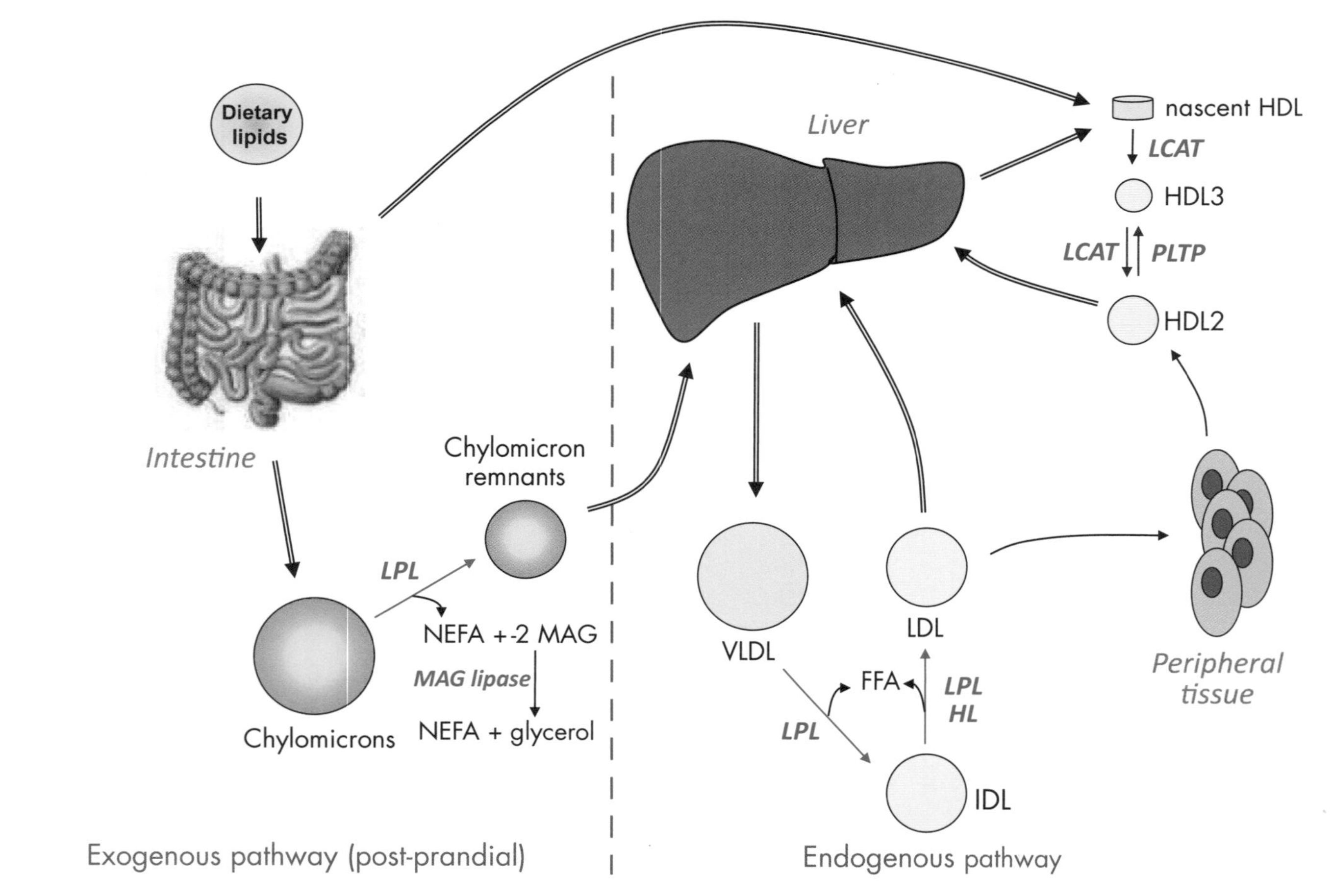

Figure 41 General metabolism of blood plasma lipoproteins making the link between the intestine, liver and peripheral tissues.

Dietary lipids are absorbed into enterocytes after cleavage by triacyl-glycerol (TAG) lipase, then TAG are re-synthesized from the products, and packaged with cholesterol, phospholipids and apo-lipoprotein B48 (apo-B48) into chylomicrons (CM), which enter the circulation through the lymph. TAG in CM are hydrolyzed by the endothelium-associated lipoprotein lipase (LPL), which is activated by apo-CII, into non-esterified fatty acids (NEFA) and 2-monoacyl-glycerols (2-MAG), ultimately hydrolyzed by MAG lipase into glycerol plus NEFA. Resulting CM remnants are rapidly cleared by the liver. Intra-hepatic TAG are assembled to apo-B100 to form nascent VLDL. Once in the circulation, TAG in VLDL are hydrolyzed by LPL to generate IDL. IDL can undergo further catabolism by LPL or hepatic lipase (HL), with loss of TAG, to produce LDL, a cholesterol-rich particle containing apo-B100. LDL can be degraded in the liver or be endocytosed by specific LDL receptors in peripheral tissues. Discoidal, nascent HDL are formed by the assembly of apo-AI, secreted by the liver and intestine, phospholipids, free cholesterol. These nascent HDL may then interact with the ATP-binding cassette transporter (ABCA1) at peripheral tissues for the reverse transport of cholesterol. HDL are also generated from TAG-rich lipoproteins (chylomicrons and VLDL) that undergo hydrolysis by LPL. Through the action of lecithin cholesterol acyl transferase (LCAT), discoidal particles become spherical because resulting cholesteryl-esters (CE) that are more hydrophobic than cholesterol are translocated to the core of the particles (HDL3). Furthermore, small, dense, CE-poor and protein-rich HDL3 particles are converted to large, light, CE-rich HDL2 particles via LCAT again. HDL CE are transported to the liver through the uptake of HDL via the scavenger receptor SR-BI.

other IDL particles can be further hydrolyzed by hepatic lipase (HL) leading to TAG degradation and formation of CE enriched particles, called LDL.

LDL are the main carrier of circulating cholesterol within the body, used by extra-hepatic cells for cell membrane building and steroid hormone synthesis. When cells require cholesterol, they increase LDL receptor synthesis, and their insertion into the plasma membrane, thus LDL particles are taken up by LDL receptors in the liver and peripheral tissues. Absorption occurs by endocytosis and the internalized LDL particles traffic through early and late endosomes to reach lysosomes, the place for CE hydrolysis, releasing mainly cholesterol.

Reverse cholesterol transport (RCT)

This transport refers to a multi-step process by which excess cholesterol is removed from tissues and returned to the liver and steroidogenic organs. HDL and apo-AI are major actors of RCT, as well as enzymes such as lecithin-cholesterol acyltransferase (LCAT), phospholipid transfer protein (PLTP), hepatic lipase (HL), and cholesteryl ester transfer protein (CETP). First of all, cholesterol is removed from cells via the ATP-binding cassette transporter A1 (ABCA1) or G1 (ABCG1), and collected by nascent/discoid HDL or apo-AI, respectively. Within HDL, LCAT catalyzes the transfer of fatty acyls from PC (lecithin) to cholesterol to produce CE. Then, CE accumulate into the core of the particle, thereby allowing the uptake of additional free cholesterol onto the HDL3 surface.

HDL increase in size as they circulate through the blood stream and incorporate more cholesterol and PL molecules from cells and other lipoproteins, by interacting with PLTP and CETP. These exchange proteins belong to a family of lipid transfer/lipopolysaccharide binding protein. Plasma PLTP mediates transfer of phospholipids from apo-B-containing-triglyceride-rich lipoprotein into HDL2, and exchanges phospholipids between lipoproteins. PLTP also is a non-specific lipid transfer protein, as it is capable of transferring all common phospholipids (PC, PE, PG, SM), DAG, α-tocopherol, cerebrosides, and lipopolysaccharides. CETP catalyzes the transfer of TAG in exchange for CE from VLDL to HDL and/or LDL, resulting in larger, relatively TAG-enriched HDL2 and LDL species. HDL2 can then follow different pathways: either uptake by the liver of apo-lipoprotein-rich HDL through LDL receptors, selective uptake of HDL-CE by the liver or other tissues involving scavenger receptor B1, or transfer to TAG-rich lipoproteins as a result of the CETP activity and uptake by the liver (Fig. 42).

Under physiological conditions, cellular cholesterol homeostasis is ensured by a tight balance between the uptake of cholesterol via the endocytosis of LDL, the storage of CE, and the elimination of cholesterol via HDL-dependent efflux mechanisms. Under pathological conditions, as the level of LDL increases, they accumulate in blood circulation, and are oxidized in the artery wall before being taken up by macrophages that become foam cells.

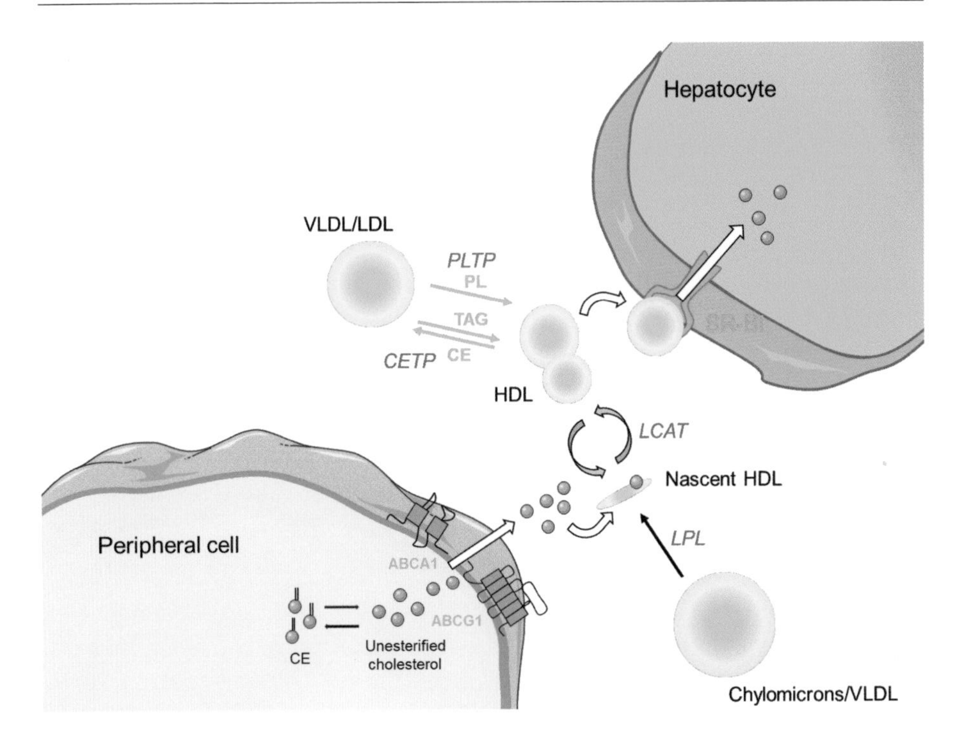

Figure 42 Reverse transport of cholesterol and metabolism within lipoproteins.

Non-esterified/unesterified cholesterol from peripheral cells is effluxed through ATP Binding Cassette (ABC) A1 or G1 to nascent HDL, secreted by the liver and intestine, or formed via the lipolysis of chylomicrons/VLDL. Cholesterol in discoidal, nascent HDL is esterified by LCAT into CE with the formation of spherical HDL (see Figure 41). Within the circulation, HDL exchange lipids with VLDL and LDL via phospholipid transfer protein (PLTP) and cholesteryl-ester transfer protein (CETP). Some CE are transferred from HDL to Apo-B containing lipoproteins (VLDL and LDL) in exchange for TAG through CETP. Phospholipids are transferred from apo-B containing lipoproteins to large HDL2 through PLTP. Stuffed-cholesterol HDL interact with specific SR-BI receptor, which is expressed in the liver, and deliver cholesterol to hepatocytes for degradation into bile acids.

This process leads to the development and progression of atherosclerosis. HDL reduce atherosclerosis, by removing cholesterol from foam cells, inhibiting LDL oxidation and limiting inflammation.

LDL and HDL exert pro- and anti-atherogenic actions, respectively. Although it is conventionally accepted that an increased plasma ratio LDL/HDL is often associated with increased cardiovascular risk, recent studies show that increasing HDL is not always enough to prevent atherosclerosis, pointing out that functional properties of HDL must be considered when choosing a therapeutic strategy to reduce the residual cardiovascular risk.

5

Conclusion and perspectives

This booklet does not cover the whole field of lipids, but focuses on those having well-characterized biological functions, especially structural lipids and lipid mediators, although we may anticipate that all lipid molecules are supposed to have their own function. This is what we believe for all the biomolecules. However, we tried to point out the structure function relationships without any attempt to be exhaustive. The contrary would have resulted in a catalog that could partly be found out of this presentation.

For most lipid species presented, we have privileged the structural aspects with more than thirty figures detailing the molecular metabolism of many classes of lipids that are well-known for their biological activities. We then focused on lipid mediators, some with hormonal functions and others that are named autacoids because their life span are short enough to restrict their action in the vicinity of their production.

A selection of reviews for further readings, starting from 2003 (first appearance of “lipidomics” in the titles of scientific articles), will complete the information by focusing on complementary subcategories and/or specific methodologies, especially in mass spectrometry because of its primary importance in the assay of lipids.

Further reading

This booklet is a summary of the structures, metabolism, and relevant biological activities of lipids doted of clearly identified functions. It has not been written as a review for a scientific journal which usually contains many literature citations. Thus, further reading will allow deeper approach, in particular for functionalities of the various molecules and metabolic pathways.

Lipidomics is emerging. Lagarde M, Géloën A, Record M, Vance D, Spener F. *Biochim Biophys Acta.* 2003 Nov 15;1634(3):61.

Mediator lipidomics. *Prostaglandins Other Lipid Mediat.* Serhan CN. 2005 Sep;77(1-4):4-14. Review.

Shotgun lipidomics: electrospray ionization mass spectrometric analysis and quantitation of cellular lipidomes directly from crude extracts of biological samples. Han X, Gross RW. *Mass Spectrom Rev.* 2005 May-Jun;24(3):367-412. Review.

The challenge of brain lipidomics. *Prostaglandins Other Lipid Mediat.* Piomelli D. 2005 Sep;77(1-4):23-34. Review.

Genomic insights into mediator lipidomics. Hla T. Prostaglandins Other Lipid Mediat. 2005 Sep;77(1-4):197-209. Review.

Cellular lipidomics. van Meer G. *EMBO J.* 2005 Sep 21;24(18):3159-65. Review.

Lipidomics and lipid profiling in metabolomics. German JB, Gillies LA, Smilowitz JT, Zivkovic AM, Watkins SM. *Curr Opin Lipidol.* 2007 Feb;18(1):66-71. Review.

The lipid maps initiative in lipidomics. Schmelzer K, Fahy E, Subramaniam S, Dennis EA. *Methods Enzymol.* 2007;432:171-83.

Lipidomics: practical aspects and applications. Wolf C, Quinn PJ. *Prog Lipid Res.* 2008 Jan;47(1):15-36. Review.

Lipidomics as a tool for the study of lipoprotein metabolism. Kontush A, Chapman MJ. *Curr Atheroscler Rep.* 2010 May;12(3):194-201.

Functional lipidomics of oxidized products from polyunsaturated fatty acids. Guichardant M, Chen P, Liu M, Calzada C, Colas R, Véricel E, Lagarde M. *Chem Phys Lipids*. 2011 Sep;164(6):544-8. Review.

Imaging mass spectrometry for lipidomics. Goto-Inoue N, Hayasaka T, Zaima N, Setou M. *Biochim Biophys Acta*. 2011 Nov;1811(11):961-9. Review.

Lipid classification, structures and tools. Fahy E, Cotter D, Sud M, Subramaniam S. *Biochim Biophys Acta*. 2011 Nov;1811(11):637-47. Review.

Applications of mass spectrometry to lipids and membranes. Harkewicz R, Dennis EA. *Annu Rev Biochem*. 2011;80:301-25. Review.

Expanding the horizons of lipidomics. Towards fluxolipidomics. Lagarde M, Bernoud-Hubac N, Guichardant M. *Mol Membr Biol*. 2012 Nov;29(7):222-8. Review.

Lipidomics applications in health, disease and nutrition research. Murphy SA, Nicolaou A. *Mol Nutr Food Res*. 2013 Aug;57(8):1336-46.

Lipidomics of essential fatty acids and oxygenated metabolite. Lagarde M, Bernoud-Hubac N, Calzada C, Véricel E, Guichardant M. *Mol Nutr Food Res*. 2013 Aug;57(8):1347-58.

Functional Fluxolipidomics of polyunsaturated fatty acids and oxygenated metabolites in the blood vessel compartment. Lagarde M, Calzada C, Jouvène C, Bernoud-Hubac N, Létisse M, Guichardant M, Véricel E. *Prog Lipid Res* 2015 Oct; 60:41-9. Review.

In addition to these specific articles, special issues are regularly published by the well-known journal *Biochimica et Biophysica Acta – Molecular and Cell Biology of Lipids*. The recent one (2015) on the "Oxygenated metabolism of PUFA: analysis and biological relevance", with 17 review articles, is especially relevant to lipid signaling.

Also a recent book entitled *Stress Oxydant et Antioxydants* by J. Pincemail (NutriDoc 2014) discusses the beneficial and deleterious effects of the oxidative stress, especially relating to dietary molecules.

About the authors

Michel Lagarde is a professor emeritus at INSA-Lyon (Lyon University). A former research scientist at Inserm, then director of an Inserm research unit, he taught biochemistry and molecular biology at the BioSciences Department of INSA-Lyon for 25 years. He founded the Institute for Multidisciplinary Biochemistry of Lipids (IMBL). He was scientific director of a functional lipidomics platform for 10 years. He has been president of several scientific societies dedicated to lipids (GERLI, ICBL, and ISSFAL). He is the co-author of around 500 articles, with an h-index of 47 according to the web of science.

His research concerns membrane lipids and lipid mediators generated from blood and vascular cells in response to specific activators, especially in the context of aging and diabetes mellitus. These lipid mediators are mainly oxygenated metabolites of polyunsaturated fatty acids with nutritional interest.

Michel Guinchardant is Professor of Biochemistry at INSA-Lyon. His research (Inserm UMR 1060, CarMeN Laboratory, INSA-Lyon) is mainly focused on oxygenated fatty acid, and their analysis by ultrahigh performance liquid chromatography coupled with tandem mass spectrometry.

Céline Luquain-Costaz is an Assistant Professor at INSA-Lyon (Inserm UMR 1060, CarMeN Laboratory, INSA-Lyon). Her research focuses on lipoprotein interaction with macrophages.

Nathalie Bernoud-Hubac is Professor at INSA-Lyon (Inserm UMR 1060, CarMeN Laboratory, INSA-Lyon). Her research interests include lipids in relation to function, metabolism and transport.

Catherine Calzada is a Senior Research Scientist at CNRS (Inserm UMR 1060, CarMeN Laboratory, INSA-Lyon). Her current research interests include oxidized lipids and plasma lipoproteins.

Evelyne Véricel is a Senior Research Scientist at Inserm (Inserm UMR 1060, CarMeN Laboratory, INSA-Lyon). Her research interests include lipids and lipid mediators in relation to function, mainly in blood platelets.

◀ Content